ELECTRIC MACHINERY

SECOND EDITION

Peter F. Ryff

Ryerson Polytechnic University

Prentice Hall Career and Technology
Upper Saddle River, New Jersey 07458

Library of Congress Cataloging-in-Publication Data

Ryff, Peter F.
    Electric machinery / Peter F. Ryff.—2nd ed.
      p.   cm.
    Includes bibliographical references and index.
    ISBN 0-13-475625-8
    1. Electric machinery.   I. Title.
    TK2182.R986   1994
    621.31′042—dc20

                      93-44578
                      CIP

Acquisitions Editor: Holly Hodder
Production Editors: Fred Dahl and Rose Kernan
Copy Editor: Rose Kernan
Designer: Fred Dahl
Cover Designer: Marianne Frasco
Production Coordinator: Ilene Sanford

Printed in the United States of America

10 9 8 7 6 5 4 3

ISBN 0-13-475625-8

Prentice-Hall International (UK) Limited, London
Prentice-Hall of Australia Pty. Limited, Sydney
Prentice-Hall Canada Inc., Toronto
Prentice-Hall Hispanoamericana, S.A., Mexico
Prentice-Hall of India Private Limited, New Delhi
Prentice-Hall of Japan, Inc., Tokyo
Pearson Education Asia Pte. Ltd., Singapore
Editoria Prentice-Hall do Brasil, Ltda., Rio De Janeiro

To my children
Lisa and Paula

Contents

Principles of DC Machines, 28

3

DC Generators, 51

4

DC Motors, 72

7

Transformers, 152

8

Induction Motors, 184

9

Single-Phase Motors and Special Machines, 225

Appendix A

Appendix B

Appendix C

Appendix D

Preface

The philosophy of this second edition is identical to that of the first edition—the emphasis is on the physical rather than mathematical concepts. Nearly all the subject matter of the first edition has been maintained, the various additions included are in response to requests from many users of the book.

There are a number of enhancements to the book. Some chapters have been slightly revised to facilitate better explanation and development of the material presented. Some of the additions are a section on induction motor starter methods to complete the coverage of this important machine; Chapter 9 has been extended to deal with specialty motors, including the stepper motor.

The computer programs have been rewritten in Fortran to facilitate conversion to other languages, if so desired. More than 20 new problems have been added to the book, and most of the answers to the problems at the end of the chapters are now included, rather than just the odd-numbered answers.

I would like to thank all instructors and students for their many suggestions to improve upon the first edition. Most of these comments have been implemented.

P. F. Ryff, Ph.D.

Magnetic Aspects of Machines

1.1 Introduction

Electrical machines and transformers have electric circuits and magnetic circuits interlinked through the medium of the magnetic flux. Electric currents flow through the electric circuits, which are made up of windings, and magnetic fluxes "flow" through magnetic circuits, which are made up of iron cores. Interaction between the currents and fluxes is the basis of the electromechanical energy conversion process that takes place in electric generators and motors. In transformers we think more in terms of an energy transfer.

In energy conversion devices we convert mechanical energy into electrical energy (generator action), or the reverse process takes place (motor action). In transformers the energy transfer is normally associated with a change in voltage and current levels. Thus magnetic circuits play an essential role in electric machines and transformers.

In this chapter we examine the basic principles as applied to magnetic systems. We are concerned with magnetic fields because the operation of such devices as relays, lifting magnets, electric machines, and so on, is based on the magnetic flux. In order to create high flux densities in a given region, the magnetic structures or members are composed for the most part of high-permeability materials. These materials ensure that the flux is contained within the regions where it is desired.

As we will see, the relationships in magnetic circuits are remarkably similar to those in electric circuits. Furthermore, as will soon be apparent, an invaluable aid in our study will be an equivalent electric circuit diagram or model, which for all practical purposes describes the device or system under study. The resulting calculations in any magnetic circuit depend greatly on the composition of the magnetic device. It is thus crucial to obtain a basic understanding of the magnetic principles involved.

1.2   Magnetic Fields

We recall from elementary physics that a magnetic field exists in the region around a magnet such as a bar or horseshoe magnet. With early magnetic field instruments, the direction of this field was established with the aid of a compass. The needle of a compass is merely a freely suspended magnetized steel needle. In the region surrounding a magnet, the magnetic force may be determined at various points.

The marked end of a compass needle always points to the earth's magnetic north pole. Since unlike magnetic poles attract and like poles repel, the marked end of the compass needle is really a south pole. By plotting the direction of the compass needle as it is moved slowly from the north pole to the south pole around a bar magnet, for example, a map of the magnetic field can be traced. Figure 1.1 shows such a map for many paths traced and shows that the magnetic force has a definite direction at all points. The figure also shows that the magnetic force acts along a curved line from the north pole to the south pole. These lines are called *lines of force*. The *field map*, as it is often called, should not be interpreted too literally. The lines are imaginary, but helpful in forming a picture of the nature of the field. The compass needle shows that the lines of force emerge from the north pole and reenter the south pole. Inside the magnet they close from the south pole to the north pole, thus forming closed loops. It can be seen that lines do not cross each other or merge into other magnetic lines. An excellent demonstration of the magnetic field around a magnet can be obtained by the familiar method of sprinkling iron filings on a sheet of cardboard placed over a magnet.

The magnetic field, the entire quantity of magnetic lines surrounding the magnet, is called the *magnetic flux*. The symbol to represent this flux is ϕ, and has the unit weber (Wb). The total magnetic flux from a magnet is not uniform. A better measure of the magnetic effect is the magnetic flux density, B. It is the number of magnetic lines per unit area. In formula form we can write,

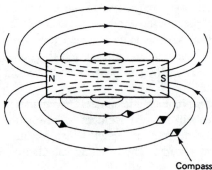

Compass
needle

FIGURE 1.1.   Magnetic field around a bar magnet.

$$B = \frac{\phi}{A} \tag{1.1}$$

The unit for B is weber per square meter (Wb/m^2), which has been given a special name, tesla (T). Thus

$$1 \, \frac{weber}{meter^2} = 1 \text{ tesla} = 1 \text{ T}$$

A is the cross-sectional area in which the flux exists, measured in square meters (m^2). Strictly speaking, B is a vector quantity, since it has magnitude as well as direction at each point. However, this distinction need not be made explicitly here, since in our study B is normally perpendicular to the area in which the flux exists, and acting along the magnetic path considered.

The magnetic fields encountered in practical machines such as motors, generators, and transformers have magnetic flux densities on the order of 1 T. The largest flux density produced in the laboratory is on the order of 10 T. Generally, fields encountered in a laboratory are on the order of 10^{-2} T (the earth's magnetic field is about 10^{-5} T).

Magnetic Field Around a Current-Carrying Conductor

When a conductor carries an electric current I (amperes), a magnetic field is created about that conductor. Electromagnets are practical applications of this phenomenon. When a compass is brought in the vicinity of a current-carrying conductor, the needle sets itself at right angles to the conductor. We associate this phenomenon with the presence of a magnetic field. When we traverse a circular path around the conductor, the needle will remain at right angles to the conductor. As we repeat this experiment at larger distances from the conductor, it shows that the field exists in concentric circles around it. Also, the farther we move away from the conductor, the weaker the field becomes (at a point near the conductor it will be more difficult to deflect the needle than at a point farther away). As the current direction in the conductor is reversed, the direction of the field is reversed. Figure 1.2 shows the magnetic field around a current-carrying conductor for current flowing into the page (indicated by $\otimes$) and for current flowing out of the page (indicated by the symbol $\odot$). The dot and cross symbols may be thought of as viewing an arrow in the direction of current flow. The dot is toward and the cross is away from the reader. To determine the direction of the field, a simple rule called Ampère's *right-hand rule* for a conductor relates the direction of current and field. It is as follows: If we grasp the conductor with our right hand with the thumb outstretched parallel to the conductor and pointing in the direction of current flow, the fingers will point in the direction of the field around the conductor. Figure 1.2c illustrates this rule.

Magnetic Field Around a Coil

To increase the magnetic field strength, we can construct a coil of many turns. Figure 1.3 illustrates a coil formed by wrapping a conductor around a tube or coil spool. It can be shown that the magnetic field strength is directly proportional to the number of turns on the coil and the current it carries. The direction of the field can be determined by Ampère's

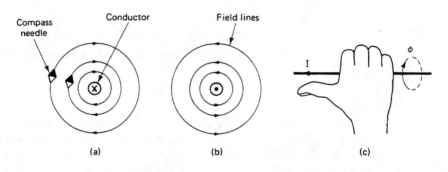

FIGURE 1.2.   Direction of magnetic field around a current-carrying conductor: (a) current directed into page; (b) out of page; (c) Ampère's right-hand rule.

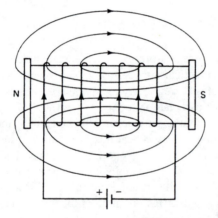

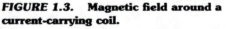

FIGURE 1.3.   Magnetic field around a current-carrying coil.

rule. Grasp the coil with the right hand and fingers following the direction of current flow. The outstretched thumb will then point in the direction of the north pole.

1.3   *Magnetic Relationships*

Magnetomotive Force

The ability of a coil to produce a flux is called the *magnetomotive force*, or MMF. It has the units of ampere-turns (At) and is given the symbol $\mathcal{F}$. Since turns is a dimensionless quantity, strictly speaking, the unit of MMF should be amperes. However, because of established practices, we will use ampere-turns (At) for MMF. Thus we can write,

$$\mathcal{F} = N I \quad \text{At} \tag{1.2}$$

where N is the number of turns on the excitation coil and I is the coil current in amperes. The magnetomotive force of a current-carrying coil corresponds to an electromotive force (EMF) in an electric circuit and may be considered a magnetic pressure, just as the EMF is considered an electric pressure. Thus, in electric circuits, EMF causes current; in magnetic circuits, MMF causes magnetic flux. Except in ferromagnetic materials, magnetic flux is strictly proportional to MMF.

EXAMPLE 1.1

A coil is uniformly wound on a wooden ring as shown in Figure 1.4. It has 600 turns, and the ring's cross section is 1 cm^2. The total magnetic flux in the ring (or *toroid*, as it is formally called) is 0.2 μWb when the coil carries a current of 800 mA. When the coil current is raised to 2 A, what will be the flux density in the toroid? Assume a linear relationship between $\mathcal{F}$ and ϕ, that is, $\phi \propto \mathcal{F}$.

Solution   Originally, $\mathcal{F} = NI = 600 \times 800 \times 10^{-3} = 480$ At is provided to create $\phi_1 = 0.2\mu$ Wb. If the current is raised to 2 A, then

$$\mathcal{F}_2 = N\,I = 600 \times 2 = 1200 \ \text{At}$$

and ϕ_2 increases to

$$\frac{\mathcal{F}_2}{\mathcal{F}_1} \times \phi_1 = \frac{1200}{480} \times 0.2 \ \mu\text{Wb} = 0.5 \ \mu \ \text{Wb}$$

The flux density in the toroid is

$$B = \frac{\phi}{A} = \frac{0.5 \times 10^{-6}}{1 \times 10^{-4}} = 0.005 \ \text{T or} \ 5 \ \text{mT}$$

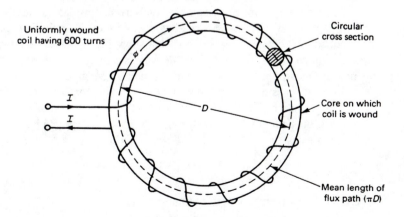

FIGURE 1.4.   Toroidal coil.

Magnetic Field Intensity

As Example 1.1 demonstrated, the magnetic flux produced is evaluated in terms of its MMF, $\mathcal{F}$. One other important quantity is the *magnetic field intensity*, H, which is the MMF per unit length along the path of the flux. According to this definition,

$$H = \frac{\mathcal{F}}{l} \quad \text{At/m} \tag{1.3}$$

The unit is ampere-turns per meter (At/m). The mean or average path length of the magnetic flux, l, is measured in meters. The magnetic field intensity is often referred to as *magnetizing force*.

Reluctance

In Example 1.1, the driving force in the circuit was increased from $\mathcal{F} = 600 \times 0.8 = 480$ At to 1200 At, which resulted in a corresponding increase in the magnetic flux, namely,

$$\frac{1200}{480} \times 0.2 \ \mu\text{Wb} = 0.5 \ \mu\text{Wb}$$

In other words, there existed a linear relationship between $\mathcal{F}$ and ϕ. We can present this as

$$\mathcal{F} = \mathcal{R}\phi \quad \text{At} \tag{1.4}$$

where $\mathcal{R}$ is the proportionality factor, normally called the *reluctance* of the magnetic circuit, expressed in At/Wb. Equation 1.4 may seem familiar: it is called *Ohm's law* for magnetic circuits. For a given coil carrying a fixed current, the flux set up in the circuit depends on the reluctance of the circuit of which it is composed. In turn, the reluctance of the medium in which the flux exists depends on its physical properties. In our case we are concerned primarily with materials that are classified as either magnetic or nonmagnetic.

Most materials are nonmagnetic, such as air, insulators, cardboard, wood, plastic, bronze, and so on. There is a small group of metals, including the ferrous irons and steel, cobalt, nickel and alloys thereof, that show strong magnetic effects. For a given coil current a much larger magnetic field density is produced in these materials than those of the nonferrous group. We will persue this in more detail shortly.

In our discussions we will not make any distinction between the magnetic properties of air and vacuum or free space. This leads to the expression of reluctance for a magnetic circuit (i.e., the path in which the flux exists), which is expressed in its physical dimensions and properties as,

$$\mathcal{R} = \frac{l}{\mu A} \quad \text{At/Wb} \tag{1.5}$$

where l = the average length of magnetic path, m

      A = the cross-sectional area of that path, m^2

      μ = permeability of the magnetic circuit material, henries/meter (H/m)

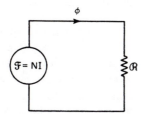

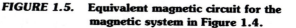

FIGURE 1.5. **Equivalent magnetic circuit for the magnetic system in Figure 1.4.**

The permeability of free space and for our purposes that of the nonmagnetic materials is a constant and is denoted by $\mu_o = 4\pi \times 10^{-7}$ H/m.

Thus the reluctance of a magnetic circuit is analogous to the resistance in an electric circuit. It is directly proportional to the length of the magnetic path and inversely proportional to its cross-sectional area. Thus $\mathcal{F} = \phi \mathcal{R}$ can be equated to Ohm's law for magnetic circuits. This leads to an equivalent magnetic circuit for Figure 1.4, analogous to an electric circuit, as shown in Figure 1.5.

In addition, reluctances may be added in series or placed in parallel. The excitation required for the core flux may be provided by multiple coils connected in series or parallel combinations. As noted, the reluctance $\mathcal{R}$ is inversely proportional to the material permeability, μ. The significance of this becomes apparent when selecting materials. For nonmagnetic materials $\mathcal{R}$ becomes relatively large, while for magnetic materials μ is large and $\mathcal{R}$ becomes correspondingly smaller.

EXAMPLE 1.2

For the toroidal core in Example 1.1 with $I=2$ A, determine
a. the reluctance of the magnetic flux path,
b. the mean circumference of the flux path, and
c. the magnetic field intensity in the core.

Solution
a. For the nonmagnetic path,

$$\mathcal{R} = \frac{l}{\mu_o A} \quad \text{or alternatively,} \quad \mathcal{R} = \frac{\mathcal{F}}{\phi}$$

Therefore,

$$\mathcal{R} = \frac{600 \times 2}{0.5 \times 10^{-6}} = 2400 \times 10^{6} \quad \text{At/Wb}$$

b.

c. $\quad l = \mu A \mathcal{R} = 4\pi \times 10^{-7} \times 1 \times 10^{-4} \times 2400 \times 10^{6} = 0.3\text{m} = 30\text{cm}$

$$\mathcal{F} = H l, \quad H = 1200/0.3 = 4800 \quad \text{At/m}$$

Permeability

The group of materials we classified as nonmagnetic all show a linear relationship between the flux density B and coil current I. We say that they have *constant permeability*. For magnetic materials, a much larger value of B is produced in these materials than in free space; therefore, the permeability of magnetic materials is much higher than μ_o. However, as it turns out, the permeability is not linear anymore but does depend on the current over a wide range. Thus the permeability is the property of a medium that determines its magnetic characteristics. In electrical machines and electromagnetic devices a somewhat linear relationship between B and I is desired, which normally is approached by limiting the current. Materials that are used and can be so linearized are known as *soft ferromagnetic materials*.

What makes these materials magnetically so different? The different behavior in iron, for example, is due to the uncompensated spin of the electrons within the iron atom. Without going into too much detail or getting too complicated, we may "visualize" it as if the iron contained many current loops which behave as molecular magnets. The iron crystals are divided into small volumes, called *domains*, which are less than 1 mm in size. As a result of the strong molecular field there is complete alignment of all these molecular "magnets" in each domain. This gives the domain itself a definite magnetic axis, although the domains are lined up randomly (see Figure 1.6). The reason for all this is that the overall system tends towards a minimum-energy configuration. The application of an external field causes the various domains' axes to line up with this field so that the resultant field becomes much more intense. Ultimately, when all the domains are lined up with the external applied field, the magnetic material is said to be *saturated*. For pure iron, B_{max} has a saturation value of about 2 T. A typical saturation curve for a magnetic material is shown in Figure 1.7, indicating the nonlinear relationship between B and H ($=NI/l \propto I$).

The alignment of the domains can be thought of as taking place in three definite stages. At first the domains closest to the applied external field will line up at the expense of others. For moderate values of applied H, indicated by the linear portion of the curve, successive rotations of other domains take place. The final stage of the curve is slow rising, due to further rotation of all the domain axes together. In the end, when saturation is reached, all are in line with the external field. Within the linear part of the curve up to the *knee* (where the curvature sets in) the B–H curve may be described by the relation

$$B = \mu_o \, \mu_r \, H = \mu \, H \tag{1.6}$$

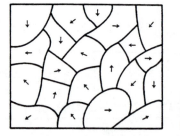

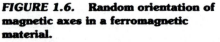

FIGURE 1.6.   Random orientation of magnetic axes in a ferromagnetic material.

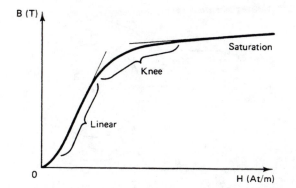

FIGURE 1.7. Typical magnetization curve for common magnetic material.

where $\mu = \mu_o \mu_r$, and μ_r is known as the *relative permeability* of the material. In our work the current in the energizing coil is dc current. The appropriate value for the magnetizing field intensity H and the corresponding flux density B, is obtained from the B–H curve for the specific core material. The permeability is then simply $\mu = B/H$.

Equation 1.6 follows by considering the relationships discussed so far, namely, $B = \phi/A$. But $\phi = \mathcal{F}/\mathcal{R}$; therefore we can write $B = (\mathcal{F}/\mathcal{R})/A$. By substituting $\mathcal{R} = l/(\mu A)$, the desired result is obtained, namely,

$$B = \mu \frac{\mathcal{F}}{l} = \mu H.$$

Note that B is a vector quantity and μ is a scalar quantity, therefore, H too is a vector quantity, having the same direction as B.

Hysteresis

Starting with an initially unmagnetized ferromagnetic core, let us apply an increasing H field to it, for example, in the toroidal core of Figure 1.8. The magnetization will increase following the initial magnetization curve until saturation is reached (see Figure 1.9). If the current is now decreased in the magnetizing coil, thereby decreasing the magnetizing force H, the initial curve will not be retraced. Rather, B will decrease along a curve, indicating there is a lag or delay in the reversal of domains.

Consequently, when H is zero ($I = 0$) there is a residual value of flux density in the core the magnitude of which, B_r, depends on the material. Thus, in effect, we have created a permanent magnet. To reduce the flux density B to zero requires a coercive field intensity $-H_c$. This implies a reversed H field, which in turn requires a reversed current flow in the coil. Increasing the current in the negative direction still further will result in a saturation level at which point all domains can be considered aligned in the opposite direction. If the applied current is periodically reversed in this manner, or alternatively, if an alternating current of low frequency is applied to the excitation coil, the hysteresis loop shown in Figure 1.9 will be traced out. It shows the multivalued and nonlinear nature of magnetization in ferromagnetic materials.

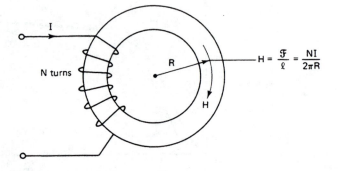

$$H = \frac{\mathcal{F}}{\ell} = \frac{NI}{2\pi R}$$

FIGURE 1.8.   **Sample magnetic core to determine magnetization curves.**

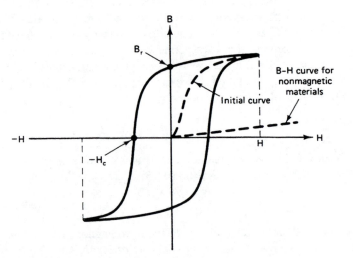

FIGURE 1.9.   Hysteresis loop for a typical ferromagnetic core.

The area of the hysteresis loop is a measure of heat energy lost per cycle. This is of great practical importance, since this generated heat must be dissipated by a magnetic material in an alternating field. Furthermore, the efficiencies of generators, motors, and transformers would be increased in the absence of hysteresis. On the other hand, on some machines it is essential to have some remanent magnetism to permit the self-excitation process to start, as we see later. It should be emphasized that in order to obtain the closed B–H loop shown in Figure 1.9, the change of field from H to $-H$ and back must be unidirectional.

In most engineering applications it is sufficient to use a *normalized B–H* curve, which is the magnetizing characteristic of the material represented by a single curve. This

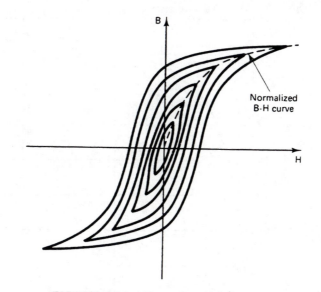

FIGURE 1.10. Normalized B–H curve.

is obtained by taking a family of hysteresis loops for several values of H. As Figure 1.10 indicates, the normalized $B–H$ curve nearly follows the locus of B for a series of loops taken up to various values of peak flux densities.

Typical magnetization curves for commonly used magnetic materials on electrical machines are shown in Figure 1.11. It is interesting to note that from an engineering point of view, magnetic materials can be broadly classified into two groups, soft and hard. In soft materials such as silicon iron (3% to 4% Si content) used in electric machines and transformers, the magnetization can be changed rapidly without much friction. Their hysteresis loops are generally characterized by a tall but narrow hysteresis loop of small area (see Figure 1.12). Materials used in such applications should have H_c relatively small. Soft materials have a uniform structure (i.e., well-aligned crystal grains). On the other hand, materials used for permanent magnets and magnetic recording tapes should have entirely different properties. They should possess great resistance to demagnetization, and H_c should be as large as possible. Such materials are classified as *magnetically hard*. Hard materials are specifically developed ferromagnetic alloys which have $B–H$ loops as shown in Figure 1.12. An example of such a material is alnico, which is iron alloyed with aluminum, nickel, and cobalt.

1.4 *Magnetic Core Losses*
Hysteresis Loss

In dealing with the hysteresis loop it was stated that the loop area represents energy dissipated as heat in the material being cycled. Only part of the energy expanded in domain alignment and rotation is recoverable, and this can be graphically demonstrated.

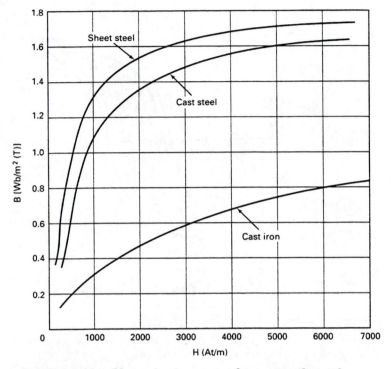

FIGURE 1.11.   Magnetization curves for commonly used magnetic materials.

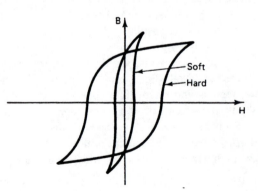

FIGURE 1.12.   B — H loops for soft and hard magnetic material.

With reference to Figure 1.13a, consider a flux change corresponding to a change in H from 0 to H. This requires a certain amount of energy as the coil current must be increased to magnetize the core. This amount of energy is represented by the area between the B–H curve and the B axis. That we may use this graphical representation of energy stored in such a manner can be confirmed by using calculus; however, we will

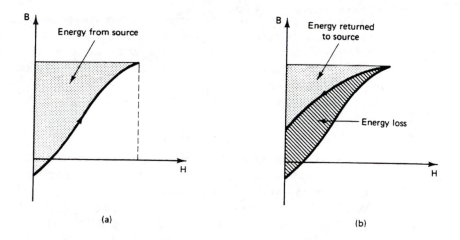

FIGURE 1.13. Hysteresis loss.

not include this analysis here. On decreasing H back to zero again, some of this energy is returned to the supply, and this too can be represented by the area between the $B-H$ curve and the B axis (see Figure 1.13b). Comparing the two areas reveals that the net energy expended is just that equal to half the area of the hysteresis loop, shown cross-hatched. Completing the cycle of magnetization will clearly require an energy represented by the whole loop. Thus magnetic materials used for electrical apparatus have a hysteresis loop area that is proportional to the power lost in that material. In fact, for a volume V of material going through f cycles of magnetization per second, the power dissipated in hysteresis loss P_h, is

$$P_h = k_h \, V \, f \, B_{max}^n \quad \text{W}$$

(1.7)

where k_h = hysteresis constant, relating loop area to B_{max}

       n = an experimentally determined constant which lies in the range of 1.5 to 2.5 depending on the material used.

       f = frequency of the supply used

   B_{max} = maximum flux density

Eddy Current Loss

In addition to hysteresis losses, there is another important loss due to the cyclic variation of the magnetic field: the eddy current loss. Since iron is a conductor, a changing flux induces EMFs and currents within the iron mass. These undesirable circulating currents, *eddy currents* as they are called, are relatively large since they encounter very little resistance in the solid iron. Subsequently, they produce power losses with corresponding heating effects. It is therefore customary in practice to use thin laminations to build up a core, so that the resistance to eddy current flow is significantly increased.

The plane of the laminations is parallel to the flux, thus restricting the eddy currents to paths of small cross section and correspondingly high resistance. The laminations are insulated from one another by a thin layer of paper or varnish. In general, the thinner the laminations, the lower the losses, since the eddy current losses are proportional to t^2, where t is the lamination thickness.

An empirical equation for the eddy current loss, P_e, is

$$P_e = k_e \, (B_{max} \, t \, f)^2 \quad \text{W/m}^3 \tag{1.8}$$

where k_e is a constant determined by the material, the other parameters being defined earlier. Equation 1.8 shows that the eddy current losses vary as the square of the frequency whereas the hysteresis losses vary directly with frequency.

Note that the eddy currents always tend to flow perpendicular to the flux and in such a direction as to oppose any changes in the magnetic field. This fact is a direct consequence of Lenz's law. Figure 1.14 shows in principle the effectiveness in limiting eddy current by laminating the core. Since Eq. 1.8 shows that the eddy current losses increase with frequency squared, they play an important role in degrading the high-frequency response of a magnetic system. Eddy currents are not always unwanted or minimized as much as economically possible. There are devices in which deliberate use is made of eddy currents. Such applications are, for example, eddy current brakes, and automobile speedometers (drag cup tachometers).

Taken together, hysteresis plus eddy current losses constitute what is generally referred to as *core losses* of magnetic systems that involve varying fluxes. In a given sample of iron it is possible to measure the iron loss. To separate the hysteresis and eddy current loss component would require two tests at different frequencies. However, this separation is rarely of interest. It should be emphasized that the core loss is extremely important in practice, since it greatly affects operating temperatures, efficiencies, and ratings of magnetic devices.

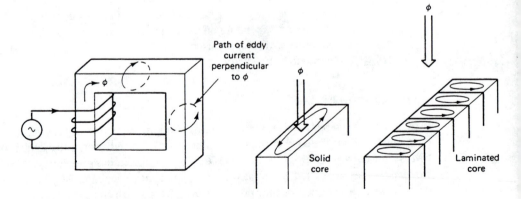

FIGURE 1.14. Laminating core to reduce eddy currents.

1.5   *Magnetic Circuit Calculations*

At this point of our study we will examine some common magnetic circuits used in practice. We must keep things simple, since it is the principles underlying these calculations that are important to us. Basically, there are two kinds of magnetic circuits, series and parallel. Analogous to electric circuits, a magnetic series circuit is one in which the flux set up by the current-carrying coil, the excitation, is common throughout the entire circuit. Parallel circuits are defined as those having more than one path for the flux to close.

Figure 1.15 illustrates this concept. Magnetic circuits may be composed of sections that have varying physical dimensions as well as being made up of different materials. Also, air gaps may be present, which will have a significant influence on the excitation requirements. Air gaps may be introduced on purpose, such as the air gap in rotating machines to separate the rotating member from the stationary one. Sometimes they are not wanted but are inevitable. For instance, in transformer construction of the core, every effort is made to reduce air gaps. In transformer construction the coil must be assembled on the core, necessitating a segmented construction. The core is reassembled after placement of the coil on the core, which is then bolted together. Typically, one desires to determine the required MMF or flux density for a specific condition. We have seen that magnetic circuits can be represented by analogous electric circuits. Thus, methods of solution applicable to electric circuits can be extended to those for magnetic circuits. The major difference we encounter is the nonlinear characteristic of the ferrous magnetic material of which the magnetic circuit is composed. It then becomes necessary to use the $B–H$ curve for the specific material used. To illustrate the use of this curve, we use the circuit of Example 1.3.

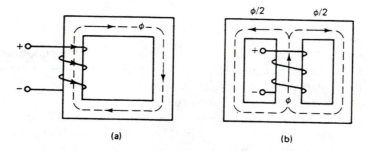

FIGURE 1.15.   Examples of magnetic circuits: (a) series; (b) parallel.

EXAMPLE 1.3

The circuit of Figure 1.16 is a magnetic core of cast steel on which a 250-turn coil is placed. For a flux of 0.48 mWb in the core, determine the necessary coil current. The core dimensions are as shown.

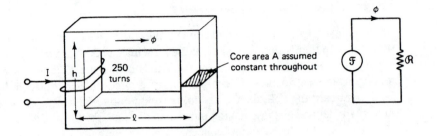

FIGURE 1.16.   Magnetic circuit for Example 1.3.   The core material is cast steel, $l = 10$ cm, $h = 8$ cm, $A = 4$ cm^2.

Solution

$$B = \frac{\phi}{A} = \frac{0.48 \times 10^{-3}}{4 \times 10^{-4}} = 1.20 \text{ T}$$

For $B = 1.20$ T, $H = 1240$ At/m (from the $B-H$ curve for cast steel of Figure 1.11). The mean or average length of the flux path is

$$l + h + l + h = 2 \times 0.1 + 2 \times 0.08 = 0.36 \text{ m}$$

Therefore, $\mathcal{F} = Hl = 1240 \times 0.36 = 446.4$ At

Hence, $I = \dfrac{\mathcal{F}}{n} = \dfrac{446.4}{250} = 1.79$ A

Fringing and Leakage Fluxes

In magnetic circuits and particularly those containing air gaps, there is a tendency for the flux to "leak out" of the magnetic path or "spread out" in the air gap. Figure 1.17 illustrates these two phenomena. As can be seen, leakage flux is not very effective since part of the flux set up by the excitation coil does not reach the air gap and therefore requires a greater magnetomotive force to compensate for this. The spreading out of the flux in the air gap is called the *fringing effect*. This is not useful either since it spreads the flux over a relatively larger area, thus in effect reducing the flux density in the air gap. The designer of electromagnetic devices must minimize these ineffective fluxes. This is why motors are made of such tight tolerances. In our calculations of magnetic circuits we will neglect these effects, because it complicates procedures considerably because of the uncertainties of the leakage flux paths and other factors.

EXAMPLE 1.4

To show the effect of an air gap in a magnetic circuit, let us assume that in the core of Example 1.3 a gap is cut of 1 mm length, as shown in Figure 1.18. What then is the required coil current to maintain the same flux?

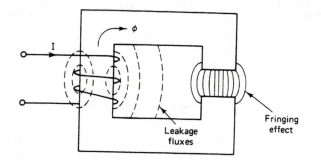

FIGURE 1.17.   Leakage fluxes and fringing effect of flux in the air gap.

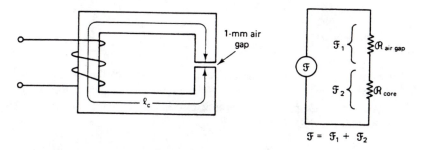

FIGURE 1.18.   Magnetic circuit and its equivalent electrical circuit for Example 1.4.

Solution $B = \phi/A = 1.2$ T and $H = 1240$ At/m as in Example 1.3. Thus the MMF drop in the iron core is $\mathcal{F}_2 = H_c\, l_c = 446.4$ At, where l_c = mean length, taken as 0.36 m. Actually, $l_c = 0.36 - 0.001 = 0.359$ m, but because of the relatively short length of the air gap as compared to the total core length, this distinction is seldom used.

To determine $\mathcal{F}_1$ we proceed as follows. The flux in the core penetrates the air gap, and if fringing effects are neglected,

$$B_g = \frac{\phi}{A} = 1.2 \text{ T} \quad \text{in the air gap.}$$

$$B_g = \mu_o H_g$$

therefore,

$$H_g = \frac{1.2}{4\pi \times 10^{-7}} = 954.9 \times 10^3 \quad \text{At/m}$$

and

$$\mathcal{F}_1 = H_g\, l_g = 954.9 \times 10^3 \times 1 \times 10^{-3} = 954.9 \text{ At}$$

The total MMF required is

$$\mathcal{F} = \mathcal{F}_1 + \mathcal{F}_2 = 954.9 + 446.4 = 1401 \; \text{At}$$

and

$$I = \frac{1401}{250} = 5.60 \; \text{A}$$

Note the relatively large MMF drop across what appears to be an insignificant length of air gap length as compared to the total length of the iron path. The current required is more than double that of the system without air gap, therefore placing more demands on the excitation source.

Furthermore, the term $H_c \, l_c$ is the magnetic potential drop across the core part of the system, and the $H_g \, l_g$ term is the air gap magnetic potential drop. Since from Eq. 1.4, $\mathcal{F} = \phi \, \mathcal{R}$, we have

$$\mathcal{F} = \phi \, \mathcal{R} = H_c \, l_c + H_g \, l_g$$

or

$$\mathcal{R} = \frac{H_c l_c}{\phi} + \frac{H_g l_g}{\phi} = \mathcal{R}_c + \mathcal{R}_g$$

which is the total reluctance of the system as seen by the source of MMF. By way of illustration of the extreme usefulness of the equivalent magnetic circuits, let us apply Kirchhoff's law, which as applied to magnetic circuits states that the applied MMF equals the sum of the MMF drops. Therefore,

$$\begin{aligned} \mathcal{F} &= \phi \mathcal{R} \\ &= \phi \, (\mathcal{R}_c + \mathcal{R}_g) \\ &= \phi \left(\frac{l_c}{\mu \, A_c} + \frac{l_g}{\mu_o \, A_g} \right) \end{aligned}$$

It illustrates that magnetic circuits can be solved equally as well by using the reluctance values of the magnetic path in question. However, care must be excercised in that the correct permeability must be obtained for the core material at the operating flux density, as described in the previous section on permeability. Examples 1.3 and 1.4 clearly indicate the significance of air gaps in magnetic systems.

Parallel Magnetic Circuits

As mentioned earlier, in parallel magnetic circuits there are multiple flux paths. We will restrict ourselves to the parallel magnetic circuit shown in Figure 1.19. The coil on the

center leg, or limb, provides the MMF NI. The flux that is produced in the center leg exists in that leg and then divides into two paths: one to the left and another to the right of the core. Assuming a symmetrical core, then the flux is distributed evenly between the two outside legs. Thus,

$$\phi_c = \phi_l + \phi_r \tag{1.9}$$

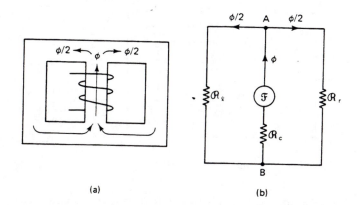

FIGURE 1.19. **Parallel magnetic circuit with analogous electric circuit.**

where

$$\phi_c = \text{flux in center leg}$$
$$\phi_l = \text{flux in left leg}$$
$$\phi_r = \text{flux in right leg}$$

Equation 1.9 is analogous to Kirchhoff's current law, but now we say that the amount of flux entering a nodal junction is equal to the amount of flux leaving that junction.

Another way of looking at it is to say that the three flux paths are in parallel, as Figure 1.19b demonstrates. This implies that the $\mathcal{F}_{A-B}$ is common to all three paths, or

$$H_l l_l = H_r l_r = \mathcal{F} - H_c l_c = \mathcal{F}_{A-B} \tag{1.10}$$

This concept will be illustrated in the next example. We should notice, however, that in the magnetic circuit the coil surrounds the center leg physically. In the equivalent electric circuit it is drawn in series with the reluctance representing the center leg.

EXAMPLE 1.5

A magnetic circuit made of sheet steel is arranged as in Figure 1.20a. The center leg has a cross-sectional area of 10 cm², and each of the side legs has a cross-sectional area of 6 cm². Calculate the MMF required to produce a flux of 1.2 mWb in the center leg, assuming magnetic leakage to be negligible.

Solution The flux density in the center leg is

$$B_c = \frac{\phi_c}{A_c} = \frac{1.2 \times 10^{-3}}{10 \times 10^{-4}} = 1.2 \text{ T}$$

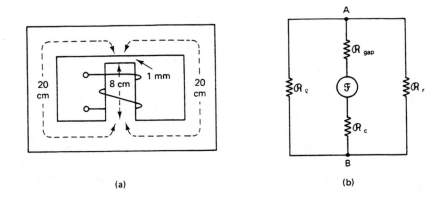

(a)

(b)

FIGURE 1.20.   **(a) Magnetic circuit for Example 1.5; (b) its electrical representation.**

From Figure 1.11, using the magnetization curve for sheet steel, $H_c = 825$ At/m. This yields

$$\mathcal{F}_c = H_c\, l_c = 825 \times 8 \times 10^{-2} = 66 \text{ At}$$

the MMF required to overcome the reluctance of the center leg. Flux density in the air gap = 1.2 T; thus, value of H_g in the air gap is

$$\frac{B}{\mu_o} = \frac{1.2}{4\,\pi \times 10^{-7}} = 0.955 \times 10^6 \ \text{At/m}$$

and

$$\mathcal{F}_{gap} = H_g\, l_g = 0.955 \times 10^6 \times 1 \times 10^{-3} = 955 \text{ At}$$

Since half the flux returns through one outer leg and half through the other, the two outer legs are magnetically identical. Therefore,

$$B_l = B_r = \frac{\phi/2}{A_r} = \frac{0.6 \times 10^{-3}}{6 \times 10^{-4}} = 1 \text{ T}$$

From Figure 1.11 the value of H for sheet steel = 500 At/m = $H_l = H_r$ and MMF for outer legs is

$$\mathcal{F}_{AB} = H_l l_l = H_r l_r = 500 \times 20 \times 10^{-2} = 100 \text{ At}$$

Hence

$$\text{total MMF } \mathcal{F} = \mathcal{F}_{AB} + \mathcal{F}_{gap} + \mathcal{F}_c = 100 + 955 + 100 = 1121 \text{ At.}$$

1.6   *Magnetic Force Between Iron Surfaces*

Consider a fixed core with an excitation coil and a movable iron core, called the *armature*. Assume the two cores have the same cross-sectional area of A m^2 and are a small distance,

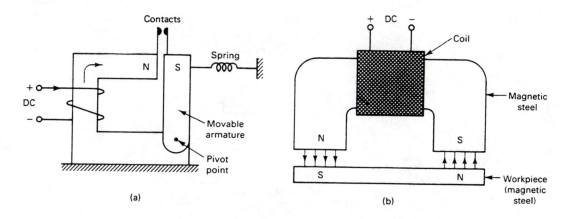

FIGURE 1.21.   (a) Relay; (b) electromagnet.

l m apart (see Figure 1.21). If the coil carries a current it will set up a flux density B in the air gap. It can be shown mathematically that an attractive force F_M will result between the two iron surfaces equal to

$$F_M = \frac{B^2 A}{2\,\mu_o} \quad \text{newtons (N)/pole face}$$

This force tends to pull the movable armature toward the fixed iron core if it is allowed to do so. Of course, we recognize this action from practical experience. The magnetic force is put to use for us in such applications as electromagnets used for lifting large sheets of metal, picking up scrap metal, iron ore separation, relays, and contactors.

The reason for this magnetic pull is readily understood, since magnetic poles of opposite polarity exist on either side of the air gap. They will therefore attract each other, and the effect of the magnetic field is such as to exert a force to close the air gap. This is the reason why a relay as shown in Figure 1.21a closes, and the electromagnet shown in Figure 1.21b can lift a steel plate (weight).

EXAMPLE 1.6

The pole faces of a lifting magnet shown in Figure 1.22 have an area of 200 cm², which is also assumed to be the cross-sectional area of the 60-cm-long flux path in the cast steel magnet. The pole faces are 30 cm apart. Determine the coil current required on the magnet if it is to lift a 600-lb cast iron plate separated by 1 mm from the pole faces.

Solution   The total force = 600 × 4.45 = 2670 N. The factor 4.45 converts pounds into newtons. Since there are 2 pole faces, the total force

$$F = 2\,\frac{B^2\,A}{2\,\mu_o} = \frac{B^2\,A}{\mu_o}$$

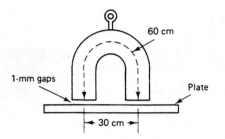

60 cm

1-mm gaps

Plate

30 cm

FIGURE 1.22.

Therefore,

$$B^2 = \frac{2670 \times 4\,\pi \times 10^{-7}}{200 \times 10^{-4}} = 0.168 \quad B = 0.41 \text{ T}$$

$$NI = H_{cs}l_{cs} + 2H_g l_g + H_{ci}l_{ci}$$

Using the appropriate curve in Figure 1.11, there results

$$NI = 215 \times 0.6 + 2\frac{0.41 \times 1 \times 10^{-3}}{4\,\pi \times 10^{-7}} + 1260 \times 0.3$$

$$= 129 + 653 + 378 = 1160 \text{ At}$$

If $N = 500$ turns, then $I = 2.32$ A.

In a practical situation we must allow for leakage fluxes, fringing, and some variation in air gap spacing due to irregularities, such as dirt, and jerks during motion depending on the mode of transport. Therefore, doubling the required current should give a satisfactory safety factor.

PROBLEMS

Note: The required magnetization curves for the various core materials are given in Figure 1.11. Leakage and fringing fluxes are not to be considered in any of the problems.

1.1 A ferromagnetic ring with a mean circumference of 36 cm and a cross-sectional area of 3 cm^2 is wound with 400 turns of wire. When the excitation current is 1.4 A, the flux is found to be 1.4 mWb. Determine the relative permeability of the iron.

1.2 A cast steel ring has a mean length of 30 cm and an air gap of 2 mm length. Find the number of ampere-turns required to produce a flux of 0.6 T in the air gap.

1.3 Find the average value of permeablity (B/H) of sheet steel for the flux densities of $B_1 = 0.4$ T, $B_2 = 0.8$ T, $B_3 = 1.2$ T and $B_4 = 1.6$ T.

1.4 Determine the relative permeabilities corresponding to the selected values in Problem 1.3. Plot these values of relative permeability as a function of the magnetic flux density. What conclusions can you draw from this curve?

1.5 A toroid of rectangular cross section as shown in Figure 1.23 has $d_1 = 12$ cm, $d_2 = 10$ cm, $t = 2$ cm, and $N = 100$ turns. If the flux in the core is 0.14 mWb, determine the required coil current.

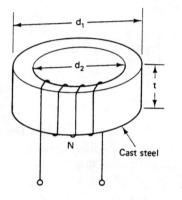

FIGURE 1.23.

1.6 If an air gap 1.2 mm length is cut in the core of Problem 1.5, by what percentage must the coil current be increased to maintain the same flux?

1.7 A cast iron pipe has a 100-turn coil wound on it, as illustrated in Figure 1.24. The dimensions (in cm) are as indicated;

a. For a coil current of 10 A, determine the flux set up in the material.
b. If a cast steel pipe had been used, what coil current would be required to establish the same flux as part in (a)?

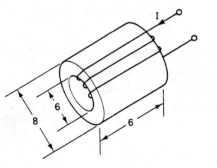

FIGURE 1.24.

1.8 Calculate the magnetization curve for: **(a)** the circuit shown in Figure 1.25 and **(b)** the same circuit with an air gap introduced as indicated by the dotted lines.

The excitation coil has 200 turns, and the material is sheet steel. Plot the total flux as a function of excitation current for flux values of 4 to 12 mWb in steps of 2 mWB.

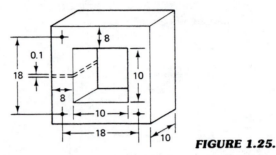

FIGURE 1.25.

1.9 Calculate the force of attraction on the armature of the cast steel electromagnet shown in Figure 1.26 when the flux density is 1.2 T in the air gap. Determine the total MMF required to produce this flux density.

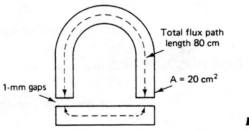

Total flux path length 80 cm

$A = 20 \text{ cm}^2$

1-mm gaps

FIGURE 1.26.

1.10 The relay shown in Figure 1.27 has a core made of cast steel and an armature of sheet steel. A flux of 0.8 T in the air gap produces a force that pulls in the armature.

   a. Determine the required current to pull in the armature.

   b. What force is exerted on the armature to overcome the spring tension?

   c. Assuming in the closed position the air gap is reduced to 0.1 mm, to what value may the current be reduced (dropout current of relay) just to retain closure? Assume that the same force is required in the closed position to overcome the spring force.

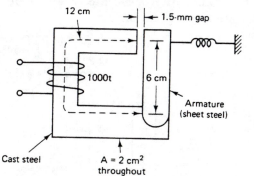

FIGURE 1.27.

1.11 The core shown in Figure 1.28 has a mean length of 26 cm and a cross-sectional area of 4 cm². The material is sheet steel. Coil *A* carries a current of 0.8 A and coil *B* a current of 0.6 A in the directions shown. Determine the flux set up in the core.

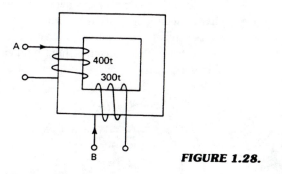

FIGURE 1.28.

1.12 Repeat Problem 1.11 with the coil current in coil *B* reversed.

1.13 A cast-steel electromagnet has an air gap of length 2 mm and an iron path of length 30 cm. Determine the number of At necessary to produce a flux density of 1.2 T in the air gap.

1.14   The core loss (hysteresis and eddy current) for a specific magnetic system is 860 W at 60 Hz. Maintaining the flux density the core loss is found to be 280 W at 25 Hz. Determine the separate hysteresis and eddy current loss at both frequencies.

1.15   A 280-turn coil is wound on the center leg of the sheet steel core shown in Figure 1.29. A flux of 1 m Wb is required in the air gap. The core is built up to a thickness of 3 cm. Determine the coil current.

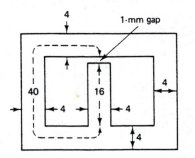

FIGURE 1.29.

1.16   For the magnetic system in Problem 1.15, a flux density of 0.8 T is desired in the outside legs. What current is required for this?

1.17   The shunt-field winding of a dc two-pole machine has 1200 turns, shown in Figure 1.30. The magnetic flux path has a net cross-sectional area of 200 cm^2. The iron portion has a mean length of 50 cm, and there are two air gaps, each 0.1 cm in length. The magnetization curve for cast steel may be taken to apply throughout the iron circuit. Determine the shunt-field current required to set up a flux of 0.02 Wb in the air gaps.

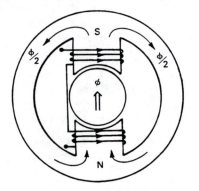

FIGURE 1.30.

1.18   The field winding of the four-pole dc machine shown in Figure 1.31 has 260 turns on each pole. The equivalent series reluctance of each magnetic path shown is

14,000 At/Wb. The air gap clearance is 1 mm, and the pole area 280 cm^2. When the field current is 1.6 A, determine the air gap flux density.

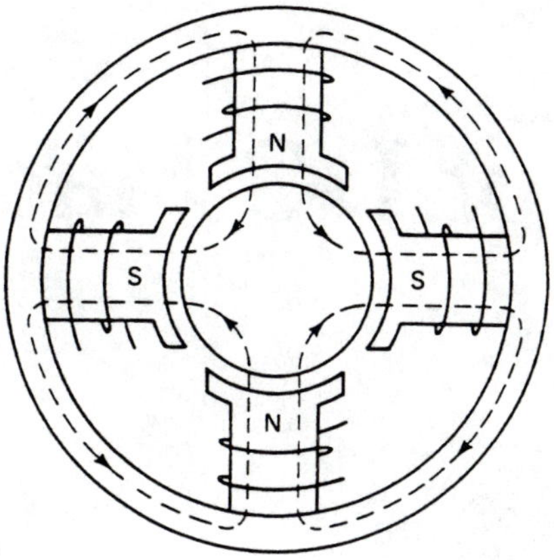

FIGURE 1.31.

Principles of DC Machines

2.1 Generator Action

Consider a conductor of length l at right angles to a uniform magnetic field of flux density B, as illustrated in Figure 2.1. The conductor can move along and make contact with a set of parallel rails. Connections are made externally to a voltmeter, to record the generated voltage between points a and b.

If the conductor moves to the right in Figure 2.1, the flux linking the circuit will change. According to Faraday's law, an EMF will be induced in the circuit. This law states: An EMF is induced in a circuit placed in a magnetic field if either (1) the flux linking the circuit changes, or (2) there is relative motion between the circuit and the magnetic field such that the conductors forming the circuit cut across the field lines. In case 1 the field varies as a function of time, which we will encounter later. Thus in our circuit as shown in Figure 2.1 a voltage e volts will be induced in the circuit because the moving conductor cuts flux. This generated voltage is recorded by the voltmeter. If the conductor moves with a relative velocity v with respect to the field, then

$$e = B\,l\,v \quad \text{V} \tag{2.1}$$

where B = magnetic flux density, T
  l = conductor length in the field, m
  v = relative velocity between field and conductor, m/s

and the voltage polarity is as indicated.

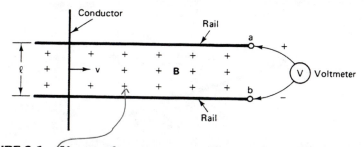

FIGURE 2.1. Linear voltage generator. The + symbol indicates magnetic field directed into page, flux density B, in tesla.

If the magnetic field is reversed (i.e., B directed out of the page) or the conductor moves from right to left, the voltage polarity also reverses. Reversing both the direction of the field B and the direction in which the conductor moves at the same time does not result in a polarity reversal. As indicated, terminal a is positive with respect to terminal b. How do we determine the terminal polarity? This is generally done by using the right-hand rule (RHR). As the name suggests, the right hand is held such that the thumb, forefinger, and middle finger are mutually perpendicular, as indicated in Figure 2.2. If the forefinger points in the direction of the field B, the thumb in the direction of conductor motion with respect to the field, the middle finger represents the direction of the generated EMF, that is, toward the positive terminal. Applying the right-hand rule to the generator in Figure 2.1 confirms the polarity shown. When the conductor motion is not perpendicular to the field, what then will be the generated EMF? Figure 2.3 illustrates this. The generated EMF turns out to be smaller than that predicted by Eq. 2.1. In fact, if we make $\theta = 90°$, there would not be any voltage generated since the conductor moves in the direction of the field and no flux is cut. This implies that Eq. 2.1 must be multiplied by some factor to account for this. As it turns out and is evident from inspection of Figure 2.3, the equation becomes

$$e = B l v \cos\theta \quad \text{V} \qquad (2.2)$$

This equation is useful if the angle between v and B changes continually, which is what happens if the conductor moves in a circle in a uniform field.

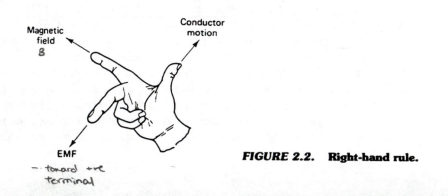

FIGURE 2.2. Right-hand rule.

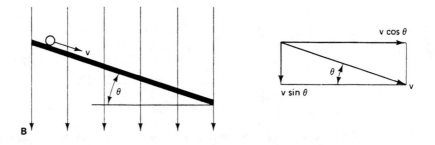

FIGURE 2.3. (a) Conductor moves at an angle with respect to magnetic field; (b) side view of Figure 2.1 with track tilted θ degrees.

2.2 Simple Generators

The linear voltage generator discussed so far is of course not very practical because of the linear motion. Commercial generators rotate about a central shaft axis so that rotary motion is involved in all instances. An elementary generator having a single turn coil is shown in Figure 2.4. As the coil rotates at a uniform angular velocity about a central axix, the generated voltage will actually be an alternating voltage. For the clockwise rotation indicated, the direction of the generated voltage in the coil side under the north pole will be directed from a to b. This is readily confirmed by using the right-hand rule.

The right-hand rule as discussed here applies to the direction of the generated voltage, which incidentally coincides with the direction of current flow inside the generator. Because the RHR determines the direction of induced EMF, it is often referred to as the *generator rule*.

Similarly, we determine the direction of the induced voltage under the south pole to be directed from c to d. When the conductor ab moves from under the north pole to the south pole, the direction of the induced EMF will reverse, so that the current will now flow from b to a. At the same time, coil side cd has moved into the north-pole region and its induced EMF is reversed, and current will flow from d to c.

Thus for one-half a revolution (for a two-pole generator) the EMF is directed around the coil from a to d; for the other half revolution, the voltage is directed around the coil from d to a. The current in the externally connected load resistor via stationary brushes in contact with a pair of slip rings A and B will be alternating current (ac).

It must be noted that this alternating current is not of sinusoidal shape, since this would only be true if the coil rotates in a uniform field. In Figure 2.4 we see that the poles are curved. This generally means that the field is more uniform at the conductor side location and the flux density more even, except for the interpole regions (i.e., the region between the poles). The result is a flat-topped induced voltage waveform more like that shown in Figure 2.5.

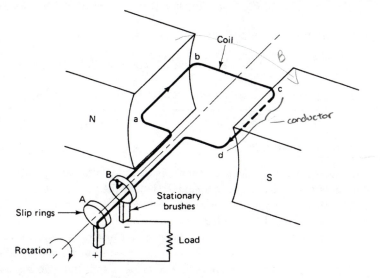

FIGURE 2.4. Elementary two-pole ac generator. Brush polarity corresponds to instantaneous voltage direction.

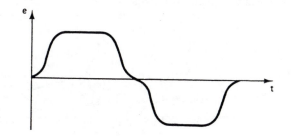

FIGURE 2.5. Generated voltage waveform in elementary generator of Figure 2.4.

EXAMPLE 2.1

For the generator in Figure 2.4, let $ab = cd = 20$ cm, the diameter of the armature 26 cm, and assume the magnetic field $B = 0.5$ T and is uniform. (For simplicity we assume the resultant waveform is sinusoidal.) For a peripheral speed of the coil sides of 12 m/s, determine the induced voltage in the coil.

Solution   For the position shown, the induced voltage in the coil side is maximum and is given by Eq. 2.1. Since both sides of the coil contribute to the generated voltage,

$$E_{max} = 2B\,l\,v = 2 \times 0.5 \times 0.20 \times 12 = 2.4 \text{ V}$$

The generated coil voltage as a functon of position is, then,

$$e_{coil} = E_{max} \cos\theta = 2.4 \cos\theta \text{ V}$$

where $\theta = 0$ is the position of the coil, as shown in Figure 2.4.

Commutator Action

All practical rotating electric generators generate alternating EMFs, and not much can be done about this. What can be done is to rectify the internal ac generated voltage so that the brush voltage, the machine terminal voltage, is direct current (dc). The mechanism for this is a mechanical rectifier, called a *commutator*.

    Let us illustrate this commutator action first before we move on to more practical machines. Figure 2.6 has a somewhat modified arrangement as compared to Figure 2.4. One slip ring is deleted and the other slip ring is split into two segments, to which the coil is now connected. This split-slip-ring arrangement represents the commutator, naturally in a simplified form because we only have one coil on the rotating member, the *armature*. We will now investigate its action. First we notice that coil side *ab* is permanently connected to segment A of the commutator, coil side *cd* to segment B. Brushes are located so that they contact each segment on top and bottom as shown.

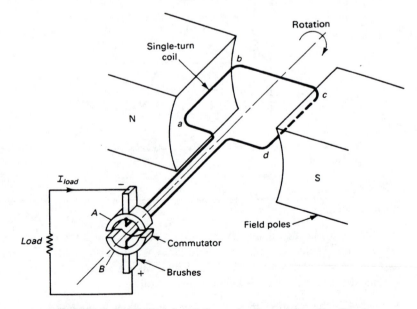

FIGURE 2.6. Principle of dc generator with split slip ring.

As the armature is rotated clockwise, the split ring rotates with it, and brushes and poles are stationary. In the position shown, the coil is located horizontally, but so is the split in the slip ring. The induced current will flow through the coil from *abcd* to the segment *B*, via the positive brush through the load to the negative brush and to segment *A*. Since the current in the external load circuit flows from the lower brush to the upper brush, they will be positive and negative, respectively. Note that as far as the generator itself is concerned, the current flow inside is from the negative to the positive terminal.

When the armature rotates so that the coil assumes a vertical position, the brushes will short-circuit both segments. This presents no problem since in that position the induced EMF in the coil is zero. Now when the armature rotates through this vertical position, coil side *ab* will enter the south pole region, and its induced voltage will reverse. Similarly, coil side *cd* will enter the north pole region, and its induced EMF will reverse as compared to the direction it was in while under the south pole. But when this happens, commutator segments *A* and *B* also have exchanged positions since they rotate in unison with the coil.

Thus, as the EMFs in the coil sides *ab* and *cd* reverse their polarity, the commutator segments to which they are connected simultaneously change positions under the stationary brushes. As a result, the polarity of the brushes remains fixed, and the current direction through the load remains the same, as shown in Figure 2.7.

This essential process of the split-slip-ring commutator to convert the internal generated ac voltage to external (the machine terminals) dc voltage is called the *commutation process*. The mechanism to achieve this, the segmented slip ring, is called the *commutator*.

In practical dc generators, many coils are properly joined together on a multisegmented commutator. This enables larger voltages to be generated, and voltage pulsations are greatly reduced. Furthermore, practical generators have four or more poles. For multipole machines, the slip ring is divided into many segments, and usually there are as many brushes as there are poles. The brushes will be interconnected by joining alternate ones to form the positive terminal while the remaining ones are joined together and form the negative terminal.

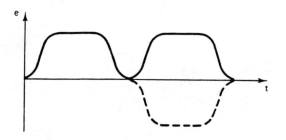

FIGURE 2.7. **Generated voltage of simple generator in Figure 2.6. Voltage is unidirectional due to commutator action.**

Armature Windings

As Figure 2.7 indicates, the generated voltage of our simple dc generator is unidirectional, but far from constant. The output voltage has a large ripple, as it is called, and as we may expect, a single turn does not produce a very large voltage. The performance can be improved considerably by adding more turns to the coil, thereby increasing the generated voltage. To decrease the voltage ripple, more coils must be added.

Consider the effect of a second coil added at right angles to the coil in Figure 2.6. Figure 2.8 shows the resulting arrangement together with the EMF *e* appearing at the terminals. Note that the coil sides are now placed in slots on the rotor (as the rotating member is called), so as to reduce the air gap length in the magnetic circuit. There is an improvement in the output waveform, but each coil is now used only half the time. The ends of each armature coil are connected to the four-segmented commutator as shown in Figure 2.8. Each commutator segment is insulated from the other, and the complete arrangement is called the *commutator*. As illustrated, the brushes ride on this commutator, which has a smooth surface. The commutator is a distinctive element in the construction of dc machines;

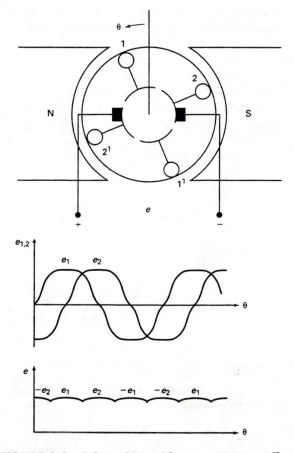

FIGURE 2.8. DC machine with two armature coils.

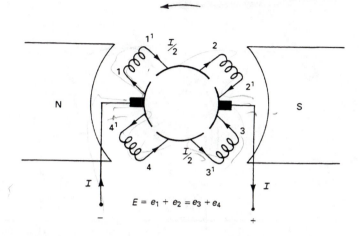

rotation

$$E = e_1 + e_2 = e_3 + e_4$$

FIGURE 2.9.   **Schematic diagram of dc machine with four coils.**

it is also the weak point of dc machines. It adds to its costs, but above all it limits the output power and speed of rotation of these machines.

In practical armature windings, the coils and commutator segments are connected so that the conductors carry current all the time, and there are usually several coils in series between the brushes. Figure 2.9 shows schematically how four coils and four commutator segments may be used to achieve this result. It can be seen that the current entering the armature divides into two parallel paths, and each path has two coils in series. The actual physical construction of the simple armature winding of Figure 2.9 is shown in Figure 2.10. As indicated, the four coils are placed in four slots. Since each coil has two sides, there are two coil sides per slot. As shown, the coils are wound so that one side of a coil occupies the bottom of a slot, while its other side is located at the top of the slot. As such, we have what is normally referred to as a *double-layer winding*. The coil connections to the commutator segments are easy to follow in this simple armature winding. The reader should compare these connections with those of Figure 2.9 to verify that they are the same. Assuming that there are three turns per coil, a particular slot will appear like that shown in Figure 2.11. The double-layer winding has important advantages. First, by placing two coil sides in one slot, the number of coils can be doubled on the armature as compared to the single layer winding of Figure 2.8. But more important, each coil will have the same form. This is utilized by preforming coils to take care of the connection crossings, prior to assembly into the machine slots. More will be said on this when discussing the ac windings (Section 5.5).

So far we have, for simplicity, concerned ourselves with the two-pole machine. More often than not, however, practical generators have four or more poles. Two-pole machines are generally used for small machines. Even in multipole machines it is interesting to note that the rectified current supplied by each coil pulsates as many times per revolution as there are poles. When multiple coils are joined in series aiding, the terminal voltage will increase. In addition, the voltage pulsations (ripple) will be less pronounced. Obviously, when there are a great many armature coils and commutator segments, the terminal voltage appearing at the brushes will approach a smooth, unvarying dc voltage.

As our generator illustrates, both coil sides of a coil are 180 electrical degrees apart. For the two-pole machine considered it means that each coil side is under an opposite pole,

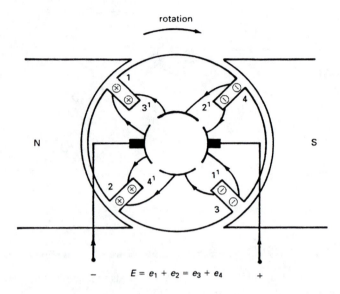

$$E = e_1 + e_2 = e_3 + e_4$$

FIGURE 2.10. **Physical construction of the armature winding in Figure 2.9.**

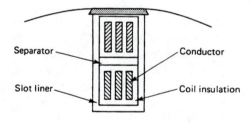

FIGURE 2.11. **Armature slot with two coil sides having three turns per coil.**

but not only that—they are at the same location under each pole. This ensures that the induced emf's in the two coil sides add to form the generated voltage in that coil.

When the dc machine has more than two poles, the armature coils still have their coil sides located at a distance of 180 electrical degrees apart. If many more armature coils are added in slots over the entire armature surface, there will be several possible winding arrangements, depending on how the coils are connected to form the closed winding. Essentially, there are two basic possibilities. Figure 2.12a shows the general arrangement of a few coils for the *lap-winding connection*, while Fig. 2.12b illustrates the *wave-winding connection*. The distinction between the lap and wave winding is significant, but all that need be noted at this point is that for a P-pole machine, a lap winding has P parallel paths through the armature between brushes. A wave winding always has two parallel paths, regardless of the number of poles. This distinction becomes important when considering voltage and current ratings of machines in the design stage. Generally, the lap winding is the choice for higher current magnitudes, the wave winding for higher voltages.

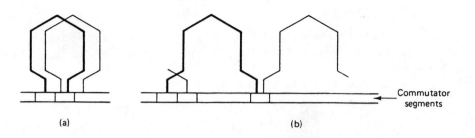

FIGURE 2.12. Possible connections of armature coils: (a) lap winding; (b) wave winding connections.

2.3 Generated Voltage

Up to now we have discussed voltage generation and the terminal voltage polarity of dc machines. The magnitude of this voltage will now be evaluated in terms of the number of armature conductors, number of poles, and the interconnection of armature conductors, that is, lap or wave wound.

Faraday's law [$e = N(d\phi/dt)$] tells us that the magnitude of the induced voltage depends on the rate at which flux is cut ($d\phi/dt$). If there are Z conductors on the armature of a P-pole machine, the total flux cut by each of the conductors is ($\phi \times P$). When the rotational speed is n r/min or $n/60$ r/s, and the Z conductors are connected in a parallel paths, the generated voltage in each conductor is ($\phi Pn/60$). With Z/a such conductors in series per parallel path, the total generated EMF becomes,

$$E_G = \left(\frac{Z P}{60 a}\right)\phi n = k_g \phi n \quad \text{V} \qquad (2.3)$$

Z = # of conductors

where $k_g = Z P/60a$ and n is the rotational speed in revolutions per minute (r/min). Since $\omega = 2\pi n/60$, Eq. 2.3 can be written in the form (consistent with the mks system of units),

$$E_G = \frac{Z P}{2\pi a} \phi \omega = k_m \phi \omega \quad \text{V} \qquad (2.4)$$

where E_G = generated EMF, V

    a = number of parallel paths on armature (determined by the type of winding, lap $a = 2$, wave $a = P$)

For a lap winding; a = P

    ω = speed of rotation, radians per second (rad/s)

    $k_m = Z P/(2\pi a)$, $V \cdot s/rad$, constant for a specific machine

showing that the generated voltage can be controlled by changing the prime-mover speed ω, or by variation of the field strength ϕ by means of the field excitation current.

EXAMPLE 2.2

A 60-kW four-pole generator has a lap winding placed in 48 armature slots, each slot containing six conductors. The pole flux is 0.08 Wb, and the speed of rotation is 1040 r/min.

a. Determine the generated voltage.

b. What is the current flowing in the armature conductors when the generator delivers full load?

Solution

a. Total number of armature conductors $Z = 48 \times 6 = 288$. For a lap winding, the number of parallel paths in the armature circuit equals the number of poles, $a = P = 4$. Thus

$$
\begin{aligned}
E_G &= \frac{ZP}{60a} \times \phi \times n \\
&= \frac{288 \times 4}{60 \times 4} \times 0.08 \times 1040 = 400 \quad \text{V}
\end{aligned}
$$

b. The total current delivered by the generator is

$$
I = \frac{P}{V} = \frac{60,000}{400} = 150 \quad \text{A}
$$

Since there are four parallel paths, each must supply 37.5 A, which is the current flowing in the conductors.

2.4   Motor Action

The function of a motor is to develop torque, which in turn can produce mechanical rotation. To develop torque, it is essential to create a force. This is done by placing the armature conductors in a magnetic field and sending a current through the conductors. In an actual motor there are many coils on the armature, like a generator. As a matter of fact, a dc generator can be run like a motor, and vice versa. Thus, in a dc motor the dc source is supplied to the brushes. The commutator will change the direction of current flow in groups of conductors under successive poles. This implies that the direction of the current in individual conductors reverses as it passes from one pole to an adjacent pole of opposite polarity. As a result, all conductors exert a force in the same direction, tending to rotate the armature.

Figure 2.13 gives us a feeling of how the force is created. When placed in a uniform field, the current-carrying conductor tends to distort that field. The conductor, in turn, will experience a force tending to move it from the higher to the lower flux density region. If the conductor current is zero, nothing will happen. In Figure 2.13c, the magnetic field density is greatest on top of the left coil side (under the north pole) and on the bottom of

Neutral
axis

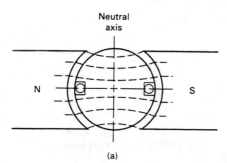

N   S

(a)

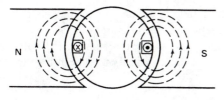

N   S

(b)

Rotation   F

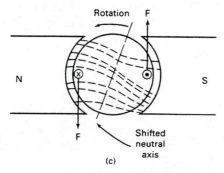

N   S

F   Shifted
neutral
axis

(c)

FIGURE 2.13. **Production of torque.
Armature will rotate counterclockwise:
(a) main field only; (b) field produced by
current-carrying armature coil;
(c) resultant fields of parts (a) and (b).**

the right coil side (under the south pole). The armature, being free to rotate on a shaft, will
therefore experience a torque that will produce counter clockwise rotation of the armature.
It is readily confirmed that by reversing either the direction of the main field or the current
in the conductor, rotation is reversed. In practice this is exactly the way to reverse the
direction of rotation of a dc motor. Note that if both the field and current direction are
changed simultaneously, rotation continues in the same direction.

    The foregoing discussion makes it clear that the force must depend on the field B and
current I in the armature conductors. Since every conductor adds to the production of total
force, it must also depend on the conductor length l; therefore,

$$F = B I l \quad \text{newtons(N)} \tag{2.5}$$

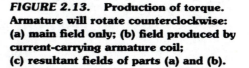

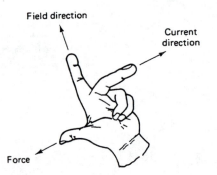

Field direction

Current
direction

Force

*FIGURE 2.14.   **Left-hand rule, or motor
rule.***

The relationship between current direction, field direction, and the developed force on the
coils may be determined by means of the left-hand rule or motor rule. As with the generator
or right-hand rule, the index finger again points in the direction of the field, the middle
finger in the direction of the current, and the thumb will then point in the direction of the
force created (see Figure 2.14).

EXAMPLE 2.3

The armature winding of a dc motor has 320 conductors, only 70% of which lie directly
under the poles, where the flux density $B = 1.1$ T. The armature diameter is 26 cm and its
length is 18 cm. The conductor current is 12 A. Determine
a. the total force created by the conductors,
b. the shaft torque developed.

Solution
a. The force created per conductor is

$$F = BIl = 1.1 \times 0.18 \times 12 = 2.376 \text{ N}$$

There are 320 conductors, but only 70% are active. The total force is

$$F = 2.376 \times 320 \times 0.70 = 532 \text{ N}$$

b. The developed torque by the armature is

$$T = F \times arm = 532 \times \frac{0.26}{2} = 69.2 \text{ N.m}$$

Counter EMF

In the preceding section motor action was discussed. The developed force on the armature
conductors causes the armature to rotate, resulting in a change of flux linkages of those

current-carrying conductors. This, we know from prior discussions, will generate an EMF in the conductors. Thus when a motor is rotating, it is acting as a generator at the same time. Obviously, the motor action is stronger than generator action because the direction of the current in the armature winding is fixed by the voltage supply. The generated EMF, however, is opposed to the impressed voltage to such an extent as to limit the armature current to the value required to drive the shaft load. Because the EMF generated directly opposes the applied voltage, in accordance with Lenz's law, it is called the *counter* or *back* EMF, E_c.

With respect to Figure 2.15, which shows two poles of a multiple-pole motor, the armature rotation will be clockwise for the assumed current direction in the armature conductors, as indicated by the crosses and dots in the circles. The direction of rotation can be verified by applicaton of the left-hand rule (motor rule).

As the armature rotates, counter EMFs are generated in the very same conductors, since they cut the main pole magnetic fluxes which are responsible for the motor action. The direction of these generated EMFs as determined by the generator rule (right-hand rule) is represented by the crosses and dots under the circles representing the current directions. Being counter EMFs, the armature current I_A must then be a function of the difference voltage of the applied line voltage V_L and the counter emf E_c. Thus by Ohm's law,

$$I_A = \frac{V_L - E_c}{R_A} \tag{2.6}$$

where R_A is the resistance of the armature winding.

In a dc motor the counter EMF limits the current to the extent dictated by the load power as we discussed and therefore must be smaller than the applied voltage. In fact, in practical machines the counter EMF will usually be in the range of 80% to 95% of the

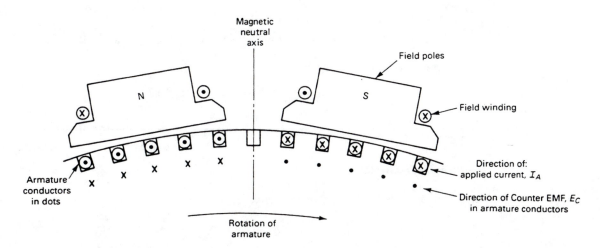

FIGURE 2.15. Direction of current flow and counter EMF in a dc motor.

terminal voltage, the higher percentage being applicable to machines of larger kilowatt (hp) ratings.

From Eq. 2.6 it is evident that the difference between the applied voltage and the counter EMF is a measure of the armature copper loss, since

$$V_L - E_c = I_A R_A \qquad (2.7)$$

When both sides of this equation are multiplied by the armature current I_A, there results,

$$V_L I_A - E_c I_A = R_A I_A^2 \qquad (2.8)$$

or, by rearranging,

$$V_L I_A - I_A^2 R_A = E_c I_A \qquad (2.9)$$

Equation 2.9 indicates that the developed power $P_d = E_c I_A$ in watts is represented by the difference of the impressed armature power $V_L I_A$ and the armature copper loss $I_A^2 R_A$. From this it is concluded that motors with a high percentage of counter EMF in terms of the applied voltage operate at a higher efficiency than those where E_c is small compared to V_L. In practice this generally translates into a design in which the designer keeps the armature resistance as low as economically feasible to keep the armature losses to a minimum.

EXAMPLE 2.4

The armature of a 120-V dc motor has a resistance of 1.5 Ω and takes 4 A when operating at full load. Calculate
a. the counter EMF produced by the armature, and
b. the developed power by the armature.

Solution
a. From Eq. 2.6,
$$E_c = V_L - I_A R_A = 120 - 4 \times 1.5 = 114 \text{ V}$$

b. The power developed is then

$$P_d = E_c I_A = 114 \times 4 = 456 \text{ W}$$

Torque Relations

The power developed internally by the motor is,

$$P_d = E_c I_A \qquad (2.10)$$

as was shown by Eq. 2.9. This is the power that must be sufficient to drive the mechanical load on the motor shaft in addition to overcoming the motor's own mechanical losses. These losses consist mainly of core losses, windage (the resistance to overcome air friction when rotating), and friction losses (occurring mainly in the bearings supporting the armature).

From mechanics we know that the torque = force × arm. In terms of our electrical machine, the force $F = BIl$ and the arm is the radius of the armature. Therefore,

$$T = BIlr \quad \text{newton-meters (N·M)} \tag{2.11}$$

It is interesting to note that both Eqs. 2.1 and 2.11 show the generated voltage and developed torque depending on the magnetic flux density. Which quantity is produced depends on whether the machine is employed as a generator or as a motor. The two electrical quantities ω and T are coupled through the medium of the magnetic field. This can be verified by dividing Eq. 2.1 by Eq. 2.11, giving

$$\frac{E}{T} = \frac{Blv}{BIlr} = \frac{\omega}{I} \tag{2.12}$$

where $\omega = v/r$ is the angular velocity of the armature. Rearranging Eq. 2.12, it states that

$$EI = \omega T = \text{Power} \tag{2.13}$$

or more explicitly, the developed electrical power equals the developed mechanical power. This statement is equally true for generator action as well as motor action. This reinforces our earlier statement that electromechanical energy conversion involves the interchange of energy between an electrical and mechanical system.

Another useful expression for motor action can be derived as follows. From Eq. 2.4, $E_G = \frac{ZP}{2\pi a}\phi\omega$, which when substituted into Eq. 2.13 and solved for the torque gives,

$$T = \frac{EI}{\omega} = \frac{ZP}{2\pi a}\phi I_A \tag{2.14}$$

or

$$T = k_m \phi I_A \tag{2.15}$$

where $k_m = \frac{ZP}{2\pi a}$ is the motor constant and I_A is the armature current. Thus the developed torque by the armature depends on the main pole field strength and the current in the armature. Note, that the machine constant k_m in the expression for torque is the same as k_g encountered for the generated voltage in Eq. 2.4, if ω is used for the rotational speed. Thus for a given machine the constant is independent on whether the machine is employed as a motor or as a generator.

EXAMPLE 2.5

The motor in Example 2.4 has a wave-connected armature winding with 360 conductors and a flux per pole of 0.01 Wb. Calculate the speed at which the motor is operating at if it has four poles.

Solution   For a wave-connected armature winding the number of parallel paths $a = 2$. The counter EMF as calculated before is $E_c = 114$ V and is a generated voltage. Therefore,

$$E_c = k_g \phi n = \frac{ZP}{60a}\phi n$$

or

$$114 = \frac{360 \times 4}{60 \times 2} \times 0.01n$$

thus,

$$n = 950 \text{ r/min}$$

2.5   Armature Reaction

When a generator supplies an electrical load or a motor a mechanical load, the armature winding will carry a current. This in turn means that a magnetic field is created by the current-carrying armature winding. This armature field will interact with the main pole field existing in the machine and set up a resulting field which is distorted. Thus armature reaction is the distorting effect of the armature flux on the value and distribution of the main pole flux.

In a way we have already encountered this in the discussion of torque production in Figure 2.13. However, that was restricted to a single coil. The idea will now be extended to a full armature winding while restricting ourselves to a two-pole machine for simplicity. Figure 2.16 shows the resultant distribution of flux due to the combination of fluxes, that is, the main pole flux combined with the resulting flux set up by the armature current. Over the leading halves of the pole faces the armature flux is in opposition to the main flux, thereby reducing the flux density. Over the trailing halves the two fluxes are in the same direction, so the flux density is increased.

Setting aside the effect of magnetic saturation, the increase in flux over one half of the pole is about the same as the decrease in the other half. Hence, in a generator the effect of armature reaction is to distort the field flux in the direction of rotation.

The immediate effect of all this is that the brushes are no longer on the magnetic neutral axis. This implies that the conductors undergoing commutation are no longer moving

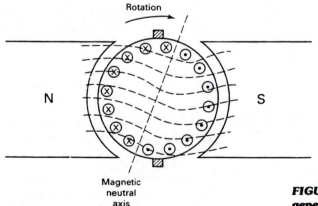

FIGURE 2.16.   Armature reaction in a dc generator.

parallel to the magnetic field; hence they have voltages induced in them. The reader may wish to refer to Figure 2.13c, where armature reaction for a motor is demonstrated. Note that for the motor the magnetic neutral axis shifts against the direction of rotation, while for the generator as illustrated in Figure 2.16, the magnetic neutral shifts in the direction of rotation.

Interpoles

At this point we may well ask the significance of a shifting magnetic neutral axis upon loading a machine. It must be remembered from earlier sections that the process of reversing the current direction in an armature coil by means of the commutator is called *commutation*. During the short time that the commutator segments, to which the armature coils are connected, are passing under the brush, the current must be completely reversed in the coil. If this reversal is not complete, an arc will occur at the commutator segment that moves under the brush. Sparking at the brushes causes the commutator to pit, increasing brush wear and commutator wear. This must be avoided. Let us examine commutation a bit further and see what is done in practical machines to overcome some of these problems.

For simplicity in considering the current reversal in a short-circuited coil, we can represent the coils and commutator segments as shown in Figure 2.17. The two ends of any coil are connected to adjacent commutator segments, just as in a lap winding. Refer to Figure 2.17, which shows a section of the commutator under a positive brush.

As the armature rotates, the segments will move over the brush. The time it takes for the brush to move across the insulation between segments 1 and 2, the current in coil b must change from I_c in one direction to I_c in the opposite direction. The process begins as soon as the brush makes contact with segment 2. Ideally, this takes place when the brushes are placed on the magnetic neutral because the coil is then not cutting any flux, and no EMF is induced in it. However, the insulated coils are placed in iron, which increases their inductance considerably. As a result, with large armature currents and short commutation times, a large EMF will be induced in the coil undergoing commutation. This induced EMF is normally referred to as a *reactance voltage*. To overcome this reactance voltage, the brushes may be moved ahead of the neutral axis in the direction of rotation for a generator and backwards in a motor. Alternatively, interpoles can be used, as is schematically illustrated in Figure 2.18. These are small additional field poles placed between the main poles. They are connected such that they alternately have the same polarity as the following main pole in the direction of rotation for a generator, or the same polarity of the trailing pole in a motor.

Thus an armature coil undergoing commutation in a generator is cutting flux which is in the same direction as that of the next main pole. The result is an induced emf in the coil that opposes the reactance voltage. Since the reactance voltage depends on the armature current, the amount of commutation flux to counteract it is made to vary. To accomplish this, the interpole winding is connected in series with the armature circuit. A machine having interpoles would then not require to have its brushes shifted every time its load changes. Figure 2.19 shows the interpoles in a six-pole dc machine. Another winding

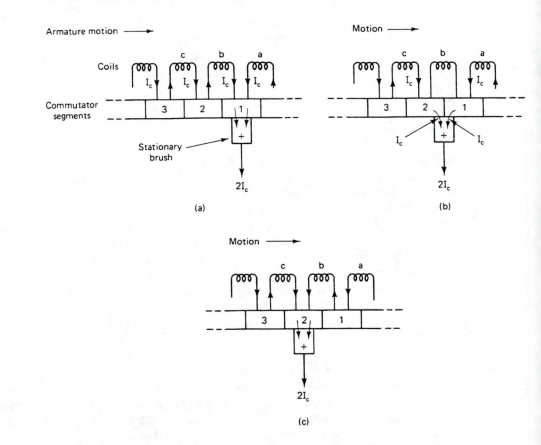

FIGURE 2.17. **Commutation process. Brush position on: (a) segment 1; (b) midway between segments 1 and 2; (c) segment 2 (note current reversal of current in coil *b*).**

that may be found on very large machines is called the *compensating winding*, also shown schematically in Figure 2.18. This winding consists of a few turns placed in the main pole faces. It also carries the armature current, and it is therefore appropriately sized to carry full load current. The resulting field produced by this winding is opposite to that produced by the current in the armature conductors that are opposite it. The compensating winding helps to counteract the effect of armature reaction. As we have discussed, armature reaction increases the flux density in one half of the main pole while decreasing it in the other half.

Because of the nonlinear behavior of the iron, the increase in flux in one pole tip is less than the decrease in the other pole tip. The net result of all this is a decreased main pole flux. Since $E_G = k_g \phi n$, the terminal voltage will decrease in a generator with increased loading. On the other hand, as we will see when discussing dc motors, a decrease in flux

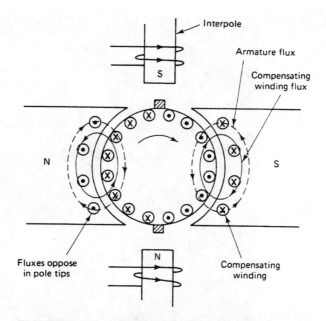

Interpole

Armature flux

Compensating winding flux

S

N

S

N

Fluxes oppose in pole tips

Compensating winding

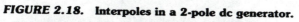

FIGURE 2.18. **Interpoles in a 2-pole dc generator.**

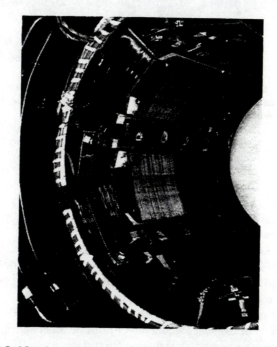

FIGURE 2.19. **Interpoles in a dc machine. (Courtesy of Siemens.)**

FIGURE 2.20.   **Dismantling of a dc machine showing the arrangements of the various parts. Note the connecting pieces of the compensating winding in particular. (Courtesy of Siemens.)**

causes a dc motor to speed up. Figure 2.20 shows the interconnections of the compensating winding in the pole faces of a large multipole dc machine.

From a user's point of view, the interpole and compensating windings appear as added armature resistance. They produce better machines and commutating characteristics, but the machine operating principles are not changed thereby.

Since every winding has resistance, heating effects inevitably show up as I^2R losses. Furthermore, there are iron losses, bearing friction, and so on—all components that tend to increase the temperature of a machine. It is for this reason that fans are normally an integral part of a machine. The generated heat must be carried away to prevent deterioration of the insulation. In environments where sparking at the brushes cannot be tolerated (sparks can start fires or explosions), the motor must be totally enclosed. In such applications the machine must be oversized, and the extra surface area will aid in cooling.

PROBLEMS

2.1 A conductor moves on, and makes contact with, a set of parallel rails at a speed of 60 cm/s. A resistor R of 10 Ω is connected to the points ab, as illustrated in Figure 2.21. The conductor length in the field is 40 cm. Determine the direction and magnitude of the current through R.

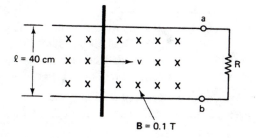

FIGURE 2.21.

2.2 In Problem 2.1, determine the direction and magnitude of the force developed on the conductor if it is moved to the right as shown.

2.3 A six-pole armature is wound with 456 conductors. The speed and flux are such that 1V is generated in each conductor. The conductor current is 40A. Determine the generated voltage and total current if the winding is

  a. lap connected,
  b. wave-connected.
  c. What is the total generated power in each case?

2.4 A conductor on the armature of a motor is 18 cm long and carries a current of 25 A. At the conductor location the magnetic field is 1T. Determine the force created on the conductor.

2.5 A two-pole motor has 260 conductors on its armature, and 70% are directly under the pole faces (i.e., active). The armature length is 16 cm, its radius 12 cm, and the field 0.8 T at the conductor locations. The armature current is 60A (this means the current in the conductors is 30A because there are 2 parallel paths). The motor is rotating at 1600 r/min. Calculate

  a. the total force developed by the armature,
  b. the mechanical power developed by the motor, and
  c. the counter EMF.

2.6 The motor in Problem 2.4 has an armature diameter of 32 cm. How many conductors are on the armature if it develops 5 hp at 600 r/min ? (1 hp = 746 W.)

2.7 A four-pole dc machine has an armature radius of 12.5 cm and an effective length of 24 cm. The poles cover 80% of the armature periphery. The armature winding is double-layered and has 33 coils, each having 6 turns. The average flux density under each pole is 0.78 T. For a lap-wound armature winding, determine

   a. the machine constant,

   b. the induced armature voltage when the armature rotates at 860 r/min,

   c. the electromagnetic torque developed when the current in the coils is 60 A, and

   d. the power developed by the armature.

2.8 Repeat Problem 2.7 for a wave-wound armature winding. Assume the current rating of the coils remains the same.

2.9 A meter movement has 120 turns wound on its movable core to which the needle is attached (see Figure 2.22). The coil measures 20 × 20 mm and is located in a field $B = 0.18$ T. The angular deflection of the coil is proportional to the coil current, 0.5 mA causes a deflection of 90°. Determine

   a. the deflection for a current of 0.25 mA, and

   b. the developed torque with a coil current of 0.5 mA.

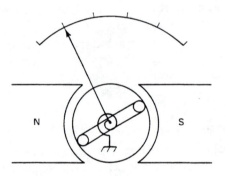

FIGURE 2.22.   Sketch of meter movement for Problem 2.9.

3

DC Generators

3.1 Introduction

For all practical purposes, the direct current generator is only used for special applications and local dc power generation. This limitation is due to the commutator required to rectify the internal generated ac voltage, thereby making large-scale dc power generators not feasible. Consequently, all electrical energy produced commercially is generated and distributed in the form of three-phase ac power. The current use of solid state converters makes conversion to dc economical. However, the operating characteristics of dc generators are still important, because most concepts can be applied to all other machines, which will be discussed in the chapters to follow.

3.2 Some Construction Details

In Chapter 2 it was established that a dc machine is capable of converting mechanical energy into electrical energy (generator) or electrical energy into mechanical energy (motor). Since generators and motors are fundamentally the same in construction, particularly for dc machines, they only differ in the way they are employed.

When the dc machine is operating as a generator, it is driven by a prime mover which rotates the rotating member, consisting of an armature core, or simply the armature, supported on a shaft. This rotating assembly is mounted in a set of bearings which are an integral part of the end shields. The armature has a winding, called the *armature winding*, placed in slots that run axially along the periphery of the armature core. Generator action takes place because of the relative motion between an existing field on the stator, in the form of field poles, and the armature conductors.

In describing the various constructional parts of a dc machine, we can divide the machine into two parts: the stationary part, generally referred to as the *stator*, and the rotating part, usually called the *rotor* (or the *armature* when referring to a dc machine). The role of the stator is to serve as the seat of the magnetic flux that is to penetrate the armature core. Except for small special machines using permanent magnets, the field circuit usually consists of a cylindrical yoke or frame to which a set of electromagnets, the field poles, are bolted.

As discussed in Chapter 2, to aid in the commutation process, dc machines may also be equipped with interpoles, thereby eliminating the need of moving the brushes (by moving the brush rigging assembly) with various loading conditions. The stator construction will then appear as illustrated in Figure 3.1. The compensating winding is another possible additional winding placed in the pole shoes, but it is only employed on very large units, as shown in Figure 2-20.

In small machines where weight is not a prime consideration, cast iron is used for the yoke frame. In larger machines cast steel is used since this material makes it possible to reduce the weight considerably and still maintain the reluctance of the magnetic circuit. It is also possible to fabricate the yoke from a rectangular steel slab by rolling it to the desired

FIGURE 3.1. **Stator of dc machine equipped with interpoles. (Courtesy of Siemens.)**

diameter and welding the ends that are butted together. End shields with their bearings and attached brush rigging assembly become part of the machine once assembled. The yoke has a base with feet or a supporting bracket upon which the entire structure rests.

The rotor, as mentioned previously, holds the insulated armature winding and is built up of a high-grade laminated steel core. A shaft through the core supports this structure as well as the commutator . The latter is located such that the brushes in the stationary brush rigging line up and rest on it. The brushes are spring loaded and have a rounded contact surface to ensure good contact with the commutator.

Field Poles

The field poles are constructed of laminated steel about 0.025 to 0.045 in. thick per lamination and having good magnetic permeability. The stack of laminations is equal to or somewhat shorter than the axial length of the armature core and are held together by rivets driven through holes in the laminations.

The shape of this assembled pole core is such that a smaller cross section exists so that the field winding or windings can be inserted on them. The pole shoe is curved and is wider than the pole core, to spread the flux more uniformly over a larger projecting area, at the same time creating a ledge upon which the field winding can have mechanical support. This construction has several desirable features, such as (1) reduced cross section of the core on which the field winding is placed permits the use of less copper; (2) the increased pole shoe area reduces the reluctance of the magnetic circuit; and (3) the winding can be assembled on the pole core prior to bolting the pole assemble to the yoke.

3.3   Field Winding Connections

The general arrangement of brushes and field windings for a four-pole machine is as shown in Figure 3.2. The four brushes ride on the commutator. The positive brushes are connected to the terminal A_1 while the negative brushes are connected to terminal A_2 of the machine. As indicated in the sketch, the brushes are positioned approximately midway under the poles. They make contact with coils that have little or no EMF induced in them, since their sides are situated between poles.

The four excitation or field poles are usually joined in series and their ends brought out to terminals marked F_1 and F_2. They are connected such that they produce north and south poles alternately.

The type of dc generator is characterized by the manner in which the field excitation is provided. In general, the method employed to connect the field and armature windings fall into the following groups (see Figure 3.3):

1. *Separately excited generators.* The field winding is connected to a separate dc supply.
2. *Self-excited generators.* They may be further clasified as:

   a. *Shunt generators.* The field winding is connected to the armature terminals.

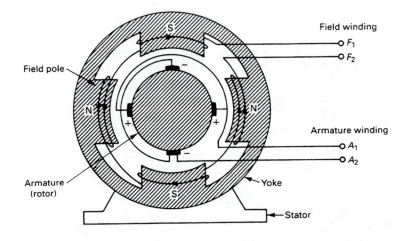

FIGURE 3.2.   Sketch of four-pole dc machine.

    b. *Series generators.* The field winding is connected in series with the armature winding.

    c. *Compound generators.* The field excitation is provided by a combination of a shunt and series field winding.

The shunt field contains many turns of relatively fine wire and carries a comparatively small current, only a few percent of rated current. The series field winding, on the other hand, has few turns of heavy wire since it is in series with the armature and therefore carries the load current.

    Before discussing the dc generator terminal characteristics, let us examine the relationship between the generated voltage and excitation current of a generator on no load. It was shown in Section 2.3 (Eq. 2.3) that the generated EMF is proportional to both the flux per pole and the speed at which the generator is driven, namely, $E_G = k_g \phi n$. By holding the speed constant it can be shown that E_G depends directly on the flux. It is not very practical to test this dependency on actual generators, because it involves a magnetic flux measurement. The flux is produced by the ampere-turns of the field coils; in turn, the flux must depend on the amount of field current flowing since the number of turns on the field winding is constant. This relationship is not linear because of magnetic saturation after the field current reaches a certain value. The variation of E_G versus the field current I_f may be shown by a curve known as the *magnetization curve* or *open-circuit characteristic*. To obtain this characteristic, a generator is driven at a constant speed, is not delivering load current, and has its field winding separately excited.

    As I_f is progressively increased from zero to a value well above rated voltage of that machine the value of E_G appearing at the machine terminals is measured. The resulting curve is shown in Figure 3.4. When $I_f=0$, that is, with the field circuit open circuited, a small voltage E_r is measured, due to residual magnetism. As the field current increases, the

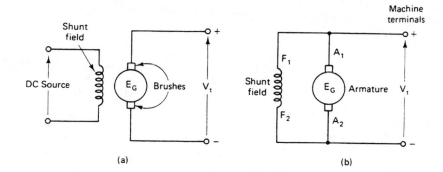

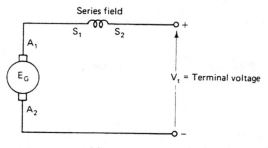

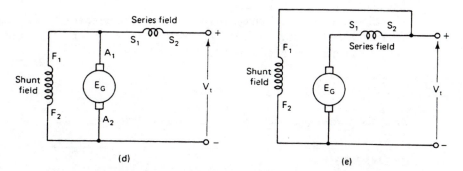

FIGURE 3.3.   **Field connections for dc generators: (a) separately excited generator; (b) self-excited, shunt generator; (c) series generator; (d) compound generator, short-shunt connection; (e) compound generator, long-shunt connecton.**

generated EMF increases linearly up to the knee of the magnetization curve. Beyond this point, increasing the field current still further causes saturation of the magnetic structure to set in. This means that a larger increase in field current is required to produce a given increase in voltage. As is apparent, the magnetic structure of a machine behaves identically

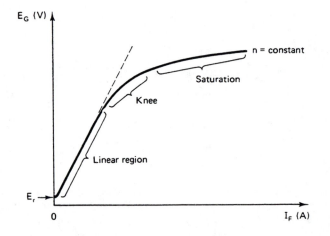

FIGURE 3.4. **Magnetization curve or open-circuit characteristic of a separately excited dc machine.**

to that discussed for magnetic materials in Section 1.3. This should come as no surprise. If measurements of E_G are taken during progressively decreasing values of I_f, a similar curve will be obtained but slightly above the existing one. This difference is due to the hysteresis (Section 1.3) property of the iron.

Since the generated voltage E_G is also directly proportional to the speed, a magnetization curve can be drawn for any other speed once the curve is determined. This merely requires an adjustment of all points on the curve according to

$$E_G' = E_G \times \frac{\omega'}{\omega} = E_G \times \frac{n'}{n} \tag{3.1}$$

where the primed quantities indicate values at the different speeds.

EXAMPLE 3.1

The open-circuit terminal voltage versus the field current (open-circuit characteristic) for a separately excited dc generator, driven at 1400 r/min, provided the following data:

E_G (V)	6	30	58	114	153	179
I_f (A)	0	0.1	0.2	0.4	0.6	0.8

Draw the no-load characteristic for 1400 r/min and 1000 r/min.

Solution   Curve (a) in Figure 3.5 shows the characteristic at 1400 r/min, obtained by plotting the data given. To obtain the characteristic at 1000 r/min, use is made of the relation $E_G = k_g \phi n$. For instance, at a field current of 0.4 A the terminal voltage is 114 V when the speed is 1400 r/min. Keeping the field current constant at this value, the open-circuit voltage at 1000 r/min becomes

$$E_G = 114 \times \frac{1000}{1400} = 81.4 \text{ V}$$

Repeating this calculation for the other given data and plotting the results gives the desired characteristic (curve b) at 1000 r/min.

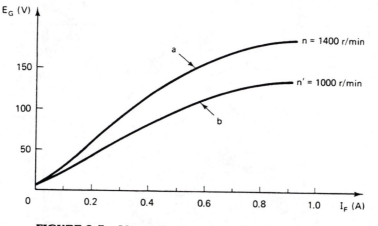

FIGURE 3.5.   Magnetization curves for Example 3.1.

3.4   *Voltage Regulation*

Let us next consider adding a load on the generator. The terminal voltage will then decrease (because the armature winding has resistance) unless some provision is made to keep it constant. A curve that shows the value of terminal voltage for various load currents is called the load or external characteristic of the generator. Figure 3.6 shows the external characteristic of a separately excited generator. The decrease in the terminal voltage is mainly due to the armature circuit resistance R_A. In general,

$$V_t = E_G - I_A R_A \tag{3.2}$$

where V_t is the terminal voltage and I_A is the armature current (or load current I_L) supplied by the generator to the load.

Another factor that contributes to the decrease in terminal voltage is the decrease in flux due to armature reaction. The armature current establishes an MMF that distorts the main field flux, resulting in a weakened flux, especially in noninterpole machines. This effect is called *armature reaction* (see Section 2.5). Figure 3.6 shows the terminal voltage

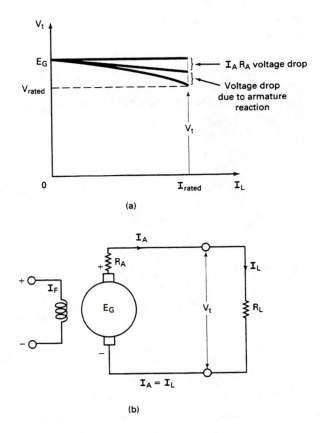

FIGURE 3.6. **(a) Load characteristic of a separately excited dc generator; (b) circuit diagram.**

versus load current characteristic. Since the iron behaves nonlinear, it gives the curve its drooping characteristic.

To have some measure by how much the terminal voltage changes from the no-load condition, it is customary to speak of *voltage regulation*. This is normally expressed as a percentage of full load voltage of the generator. Thus

$$\text{voltage regulation} = \frac{V_{NL} - V_{FL}}{V_{FL}} \times 100\% \qquad (3.3)$$

where V_{NL} is the no-load terminal voltage and V_{FL} is the full-load terminal voltage. If the voltage remains relatively constant with loading, the voltage regulation is low and is said to be good.

A separately excited generator has a low voltage regulation but has the disadvantage of requiring an external dc source. It is therefore used only where a wide range of terminal voltages is required.

EXAMPLE 3.2

The separately excited dc generator of Example 3.1 is driven at 1400 r/min, and the field current is adjusted to 0.6 A. If the armature circuit resistance is 0.28 Ω, plot the output voltage as the load current is varied from 0 to 60 A. Neglect armature reaction effects. If the full-load current is 60 A, what is the voltage regulation?

Solution   From Example 3.1, $E_G = 153$ V when the field current is 0.6 A, which is the open-circuit terminal voltage. When the generator is loaded, the terminal voltage is decreased by the internal voltage drop, namely,

$$V_t = E_G - I_A R_A$$

For a load current of, say, 40 A,

$$V_t = 153 - 40 \times 0.28 = 141.8 \text{ V}$$

Repeating this calculation for a number of load currents, the external characteristic can be plotted as shown in Figure 3.7. At full load the terminal voltage

$$V_t = 153 - 60 \times 0.28 = 136.2 \text{ V}$$

Thus the voltage regulation is

$$\frac{V_{NL} - V_{FL}}{V_{FL}} \times 100 = \frac{153 - 136.2}{136.2} \times 100 = 12.3\%$$

It should be noted that strictly speaking, the generated voltage E_G is not a constant voltage. As we have discussed before, with increasing loads on the generator, armature reaction sets in. This in turn causes the generated emf to drop somewhat. However, in our calculations we will keep things simple and neglect this effect.

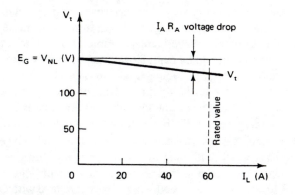

FIGURE 3.7. Calculated load characteristic of Example 3.2.

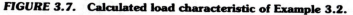

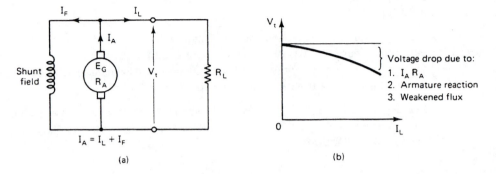

FIGURE 3.8. **Shunt generator: (a) circuit; (b) load characteristic.**

3.5 Shunt or Self-Excited Generators

A shunt generator has its shunt field winding connected in parallel with the armature so that the machine provides its own excitation, as indicated in Figure 3.8. The question arises whether the machine will generate a voltage and what determines the voltage.

For voltage to "build up"" as it is called, there must be some remanent magnetism in the field poles. Ordinarily, if the generator has been used previously there will be some remanent magnetism. We have seen in Section 3.3 that if the field is disconnected, there is a small voltage E_r generated due to this remanent magnetism, provided the generator is driven at some speed. Connecting the field for self-excitation, this small voltage will be applied to the shunt field and drive a small current through the field circuit. If this resulting small current in the shunt field is of such a direction that it weakens the residual flux, the voltage remains near zero and the terminal voltage does not build up. In this situation the weak main pole flux opposes the residual flux.

If the connection is such that the weak main pole flux aids the residual flux, the induced voltage will become larger. This in turn means more voltage applied to the main field and the terminal voltage increases rapidly to a large, constant value. The build-up process is readily seen to be cumulative; that is, more voltage increases the field current, which in turn increases the voltage, and so on. The fact that this process terminates at a finite voltage is due to the nonlinear behavior of the magnetic circuit. In steady state the generated voltage causes a field current to flow that is just sufficient to develop a flux required for the generated EMF that causes the field current to flow.

The circuit carries only dc current, so that the field current depends only on the field circuit resistance, R_f. This may consist of the field winding resistance R_F plus a field rheostat resistance R in series with it. For a given value of field circuit resistance R_f, the field current depends on the generated voltage in accordance with Ohm's law. This is illustrated in Figure 3.9 as a straight line, R_{f1}. When the field circuit resistance is increased to R_{f2},

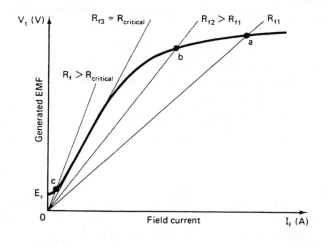

FIGURE 3.9.   Variation of E_G with field circuit resistance.

by increasing the field rheostat resistance, it changes the slope of the field resistance line and thereby the terminal voltage. Thus the open-circuit terminal voltage can be controlled by adjusting the field current.

It should be evident that on a new machine or one that has lost its residual flux because of a long idle period, some magnetism must be created. This is usually done by just connecting the field winding to a separate dc source for a few seconds. This procedure is generally known as *flashing the field*.

Increasing the field circuit resistance still further a value is reached where the field resistance line R_{f3} becomes tangent to the excitation curve (see Figure 3.9). This is a situation where the generated EMF is extremely sensitive to variations of the field circuit resistance. This value of the shunt field circuit resistance R_{f3} is known as the *critical resistance* of the shunt field circuit. Exceeding this resistance value results in failure of the generated voltage to build up. The intersection of the field circuit resistance line and the excitation curve then occurs at a low voltage value nearly equal to E_r, as indicated by point c in Figure 3.9.

Another condition that will prevent the generator voltage to build up is too low an armature speed. Thus for a given field circuit resistance there will exist a speed below which the generator will not build up. This is a direct consequence of Eq. 3.1. At lower speeds the magnetization curve is adjusted, which in turn affects the critical field circuit resistance.

Finally, there is a fourth condition which will prevent the voltage from building up. The polarity of the generated voltage depends on the direction of rotation. If a generator does not build up and all other conditions are met, the polarity of the brushes must be reversed. This can be done by reversing the direction of rotation. By reversing the direction of rotation, the polarity of the main field with respect to the residual flux will be reversed. If the voltage does build up now, it means that the main field and residual field were opposing.

EXAMPLE 3.3

The data obtained for the magnetization curve of a four-pole dc machine when run at 1500 r/min are represented by the magnetization curve given in Figure 3.10.

a. Determine the voltage to which the machine will build up as a self-excited generator if the field circuit resistance is 120 Ω.

b. Determine the critical field circuit resistance.

c. When the machine is driven at 1350 r/min [10% less then in part (a)], to what voltage would it build up?

Solution

a. Plot the field resistance line on the magnetization curve. For this select a convenient value of current, say 1 A; then the corresponding voltage is $1 \times 120 = 120$ V in accordance to Ohm's law. The voltage to which the generator builds up is the point of intersection between the magnetization curve and the field resistance line, or 102 V.

b. Draw a tangent to the magnetization curve as shown in Figure 3.10. The slope of this line is the critical field resistance at the given speed. Its value is found to be 60 V/0.28 A = 214 Ω. This value could be obtained by selecting any other convenient value of voltage and finding the corresponding current.

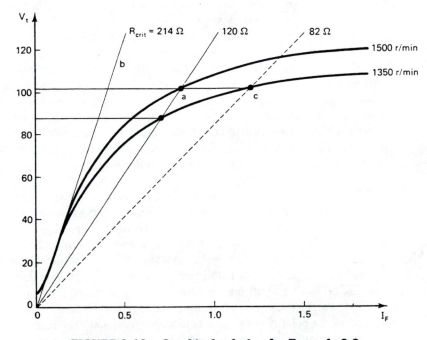

FIGURE 3.10. Graphical solution for Example 3.3.

c. The magnetization curve for a speed of 1350 r/min is shown in Figure 3.10, using the procedure outlined in Example 3.1. By inspection it is seen that the voltage builds up to 87 V. It may be interesting to observe that the field circuit resistance should be reduced to 82 Ω, in order for the terminal voltage to assume the same value as in part (a). This is indicated by point *c* in Figure 3.10.

The variation of terminal voltage with load current for a shunt generator is greater than that of the separately excited generator. This is due to the flux decreasing with increased loading, which in turn decreases the terminal voltage. Remember, the shunt field is in parallel with the armature. Figure 3.8b shows the external characteristic for the shunt generator.

EXAMPLE 3.4

A shunt generator has a field resistance of 60 Ω. When the generator delivers 6 kW, the terminal voltage is 120 V, while the generated EMF is 133 V. Determine
a. the armature circuit resistance, and
b. the generated EMF when the output is 2 kW and the terminal voltage is 135 V.

Solution
a. The circuit diagram when delivering 6 kW is as shown in Figure 3.11. The load current is

$$I_L = \frac{6,000}{120} = 50 \text{ A}$$

The field current supplied by the armature is 120/60 = 2 A. Therefore,

$$I_A = I_F + I_L = 52 \text{ A}$$

Since $V_t = E_G - I_A R_A$,

$$R_A = \frac{E_G - V_t}{I_A} = \frac{133 - 120}{52} = 0.25 \ \Omega$$

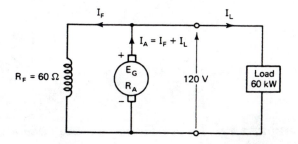

FIGURE 3.11.   Circuit diagram for the solution of Example 3.4.

b. For a load of 2 kW the terminal voltage is 135 V. Thus the load current is

$$I_L = \frac{2,000}{135} = 14.9 \text{ A}$$

and the field current is

$$I_F = \frac{135}{60} = 2.25 \text{ A}$$

Therefore,

$$I_A = 14.9 + 2.25 = 17.15 \text{ A}$$

Then

$$E_G = V_t + I_A R_A = 135 + 17.15 \times 0.25 = 139.3 \text{ V}$$

Series Generator

As mentioned previously, the field winding of a series generator is in series with the armature. Since it carries the load current the series field winding consists of only a few turns of thick wire. At no load, the generated voltage is small due to residual field flux only. When a load is added, the flux increases, so does the generated voltage. Figure 3.12 shows the load characteristic of a series generator driven at a certain speed. The dashed line indicates the generated EMF of the same machine with the armature open-circuited and the field separately excited. The difference between the two curves is simply the IR drop in the series field and armature winding, such that

$$V_t = E_G - I_A (R_A + R_s) \tag{3.4}$$

where R_s is the series field winding resistance.

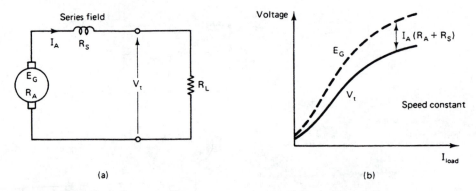

(a)   (b)

FIGURE 3.12. Series generator: (a) circuit diagram; (b) load characteristic.

Series generators are obviously not suited for applications requiring reasonably good voltage regulation. Therefore, they have been used very little and only in special applications, for example, as voltage boosters and arc welding applications. The generator is then placed in series with a supply line. When the current demand increases, the generated voltage of the machine goes up because the field current, being the load current, increases. The generated voltage thus compensates for the voltage drop in the line.

Compound Generator

The compound generator has both a shunt and a series field winding, the latter winding wound on top of the shunt winding. Figure 3.13 shows the circuit diagram. The two windings are usually connected such that their ampere-turns act in the same direction. As such the generator is said to be *cumulatively compounded*. The shunt connection illustrated in Figure 3.13 is called a *long shunt* connection. If the shunt field winding is directly connected across the armature terminals, the connection is referred to as *short shunt*. In practice the connection used is of little consequence, since the shunt field winding carries a small current as compared to the full load current. Furthermore, the number of turns on the series field winding is relatively small compared to that of the shunt field winding. This implies it has a low resistance value and the corresponding voltage drop across it at full load is minimal.

Curve *s* in Figure 3.14 represents the terminal characteristic of the shunt generator. By the addition of a small series field winding the drop in terminal voltage with increased loading is reduced as indicated. Such a generator is said to be *undercompounded*. By increasing the number of series turns, the no-load and full-load terminal voltage can be made equal; the generator is then said to be *flat compounded*. If the number of series turns is more than necessary to compensate for the voltage drop, the generator is *overcompounded*. In that case, the full-load voltage is higher than the no-load voltage. The overcompounded generator may be used in instances where the load is at some distance from the generator. The voltage drops in the feeder lines are then compensated for with increased loading.

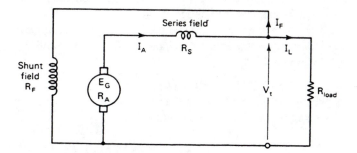

FIGURE 3.13. Compound generator.

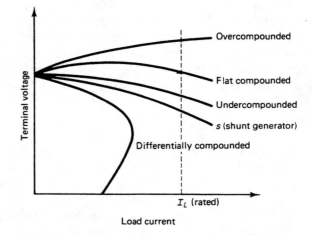

FIGURE 3.14. Terminal characteristics of compound generators compared with that of the shunt generator.

Reversing the polarity of the series field in relation to the shunt field, the fields will oppose each other more and more as the load current increases. Such a generator is said to be *differentially compounded*. It is used in applications where load conditions could occur approaching those of a short circuit. An example would be where feeder lines could break and short circuit the generator. The short-circuit current, however, is then limited to a "safe" value. The terminal characteristic for this type of generator is also shown in Figure 3.14. Compound generators are used more extensively than the other types because it may be designed to have a wide variety of terminal characteristics.

EXAMPLE 3.5

Determine the number of series turns required on each pole of a compound generator to enable it to maintain the voltage at 240 V between no load and full load of 20 kW. Without the series winding it is found that the shunt current has to be 4 A on no load and 5 A at full load to maintain the terminal voltage at 240 V. The number of turns per pole on the shunt field winding = 600.

Solution   Figure 3.15 gives the circuit diagram of the compound generator. A long shunt connection is assumed. On no load, the field excitation current is supplied by the armature. The resulting MMF of a shunt field pole is,

$$\mathcal{F} = nI = 600 \times 4 = 2400 \text{ At}$$

The small voltage drop across the series field winding will be assumed negligibly small. On full load, the shunt field current is 5 A to maintain the terminal voltage at 240 V; in other words, the required MMF is then $\mathcal{F} = nI = 600 \times 5 = 3000$ At/pole. It is the

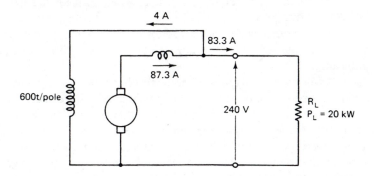

FIGURE 3.15.   Compound generator at full load for Example 3.5.

difference in MMF that would have to be supplied by the series field winding to give flat compounding while keeping the shunt field at the initial value of 4 A. Thus MMF series field is 3000-2400 = 600 At/pole. The full-load current, which also flows through the series field winding, is

$$I_{FL} = \frac{P_L}{V_t} = \frac{20,000}{240} = 83.3 \text{ A}$$

Therefore, the number of turns required on the series field to provide an MMF of 600 At is

$$\text{series field turns} = \frac{600 \text{ At}}{83.3 + 4} = 6.9/\text{pole}$$

From Example 3.5 it follows that the required number of turns on the series field winding per pole is

$$N_s = \frac{N_f \, \Delta I_f}{I_s(rated)} \tag{3.5}$$

where ΔI_f is the increase in shunt field current required to maintain the terminal voltage from no load to rated load without the series field, and I_s is the rated load or armature current flowing through the series field winding.

In Example 3.5, 6.9 series field turns/pole are needed for flat compounding. It is often desirable to provide the series field winding with a number of turns in excess of that required. This enables connection of a low-resistance shunt, known as a series field *diverter resistance*, in parallel with the series field to bypass part of the current. The degree of compounding is thereby adjustable.

In Example 3.5 it is thus advisable to provide 8 turns/pole on the series field winding. A small amount of overcompounding is the result which can be corrected by using a series field diverter. The result is that part of the armature current is diverted away from the series field winding. Consequently, the amount of current through the diverter resistance does not contribute to the series field MMF.

With 8 turns/pole on the series field winding, the series field current need only be $600/8 = 75$ A to provide the required MMF for flat compounding. Thus $87.3 - 75 = 12.3$ A must be diverted away from the series field. This means that the diverter resistance in parallel with the series field winding should be $75/12.3 = 6.10$ times the series field winding resistance.

As illustrated, the full-load terminal voltage can be maintained at the no-load value by proper degree of compounding. Other methods of voltage control are the use of rheostats, for instance, in the field circuit. However, with changing loads it requires a constant adjustment of the field rheostat to maintain the voltage. A more useful arrangement, which is now common practice, is to use an automatic voltage regulator with the generator. In essence, the voltage regulator is a feedback control system. The generator output voltage is sensed and compared to a fixed reference voltage. Any output voltage deviation from the reference voltage gives an error signal which is fed to a power amplifier. The power amplifier supplies the field excitation current. If the error signal is positive, for example, the output voltage is larger than desired and the amplifier will reduce its current drive. In doing so the error signal will be reduced to zero.

3.6   Efficiency

The *efficiency* of any machine is the ratio of the output power to the input power. The input power is provided by the prime mover to drive the generator. Because part of the energy delivered to the generator is converted into heat, it represents wasted energy. These losses are generally minimized in the design stage; however, some of these losses are unavoidable. Thus the efficiency, usually expressed as a percentage, is, by definition,

$$\text{efficiency} = \frac{\text{output power}}{\text{input power}} \times 100\% \tag{3.6}$$

or

$$\eta = \frac{\text{output power}}{\text{output power} + \text{losses}} \times 100\% \tag{3.7}$$

where η (pronounced "eta") is the symbol used to denote efficiency.

The losses of generators (and the same applies to motors) may be classified as follows:

1. *Copper losses.* The copper losses are present because of the resistance of the windings. Currents flowing through these windings create ohmic losses (i.e. $I^2 R$ losses). The windings that may be present in addition to the armature winding are the field windings, interpole, and compensating windings.

2. *Iron losses.* As the armature rotates in the magnetic field, the iron parts of the armature as well as the conductors cut the magnetic flux. Since iron is a good conductor of electricity, the EMFs induced in the iron parts causes currents to flow through these parts. These are the eddy currents we have encountered before. Laminating the core will reduce the eddy currents greatly without increasing the reluctance of the magnetic circuit since the laminations are parallel to the magnetic field. The eddy current loss is thereby minimized.

Another loss occurring in the iron is due to the hysteresis loss, as discussed in Chapter 1. Hysteresis loss is present in the armature core since the rotating armature continually moves through the alternate stationary magnetic field poles.

3. *Other rotational losses.* These are bearing friction loss, friction of the brushes riding on the commutator, and windage losses. Windage losses are those associated with overcoming air friction in setting up circulating currents of air inside the machine for cooling purposes. These losses are usually very small.

EXAMPLE 3.6

A 10-kW 125-V compound generator has rotational losses amounting to 580 W. The shunt field resistance is 62.5 Ω, the armature resistance 0.12 Ω, and the series field resistance 0.022 Ω. Calculate the full-load efficiency.

Solution   The full load current is

$$I_L = \frac{10.000}{125} = 80 \text{ A}$$

Assuming a long-shunt connection, the field current is

$$I_F = \frac{125}{62.5} = 2 \text{ A}$$

Then $I_A = 80 + 2 = 82$ A, which is the current through the series winding. The $I^2 R$ losses can now be determined:

Armature: $82^2 \times 0.12$	=	807 W
Series field: $82^2 \times 0.022$	=	148 W
Shunt field: $2^2 \times 62.5$	=	250 W
Rotational loss	=	580 W
Total loss	=	1785 W

Hence

$$\eta = \frac{P_{out}}{P_{out} + P_{loss}} \times 100 = \frac{10,000}{11,785} \times 100 = 84.9\%$$

An alternate way to solve for the efficiency is as follows:

$$\begin{aligned} E_G &= V_t + I_A (R_S + R_A) \\ &= 125 + 82 (0.022 + 0.12) = 136.6 \text{ V} \end{aligned}$$

The power developed by the armature is

$$P_d = E_G I_A = 136.6 \times 82 = 11,205 \text{ W}$$

Therefore, the power input at the shaft is $P_d + P_{rot} = 11,785$ W and

$$\eta = \frac{10,000}{11,785} \times 100 = 84.9\%$$

as above.

To determine the efficiency from measurements, we need to know the various resistance values accurately. These values are usually obtained using the volt-ampere method. In practical machines there is always a voltage drop across the brushes which depends on the current and to some extent on the state of the commutator. Although it is neglected in our calculations, it is common practice to allow a 1- to 2-V brush voltage drop.

Rotational losses can be measured to a fair degree of accuracy by running the generator as a motor at various loads. By measuring the line input quantities such as current, voltage, and power, subtracting the I^2R losses at each load condition, the rotational loss versus speed characteristic may be obtained. Note that this method measures the total rotational loss and does not separate windage, hysteresis, and eddy current loss. However, this separation is rarely required.

PROBLEMS

3.1 If the no-load voltage of a separately excited generator is 125 V at 900 r/min, what will be the voltage if the speed is increased to 1200 r/min? Assume constant field excitation.

3.2 A separately excited generator supplies 180A at 125V to a resistive load when its speed is 1200 r/min. What will the current be if the speed is reduced to 1050 r/min, the field remaining constant. The armature resistance is 0.06 Ω.

3.3 A separately excited generator has an open-circuit terminal voltage of 144 V. When loaded, the voltage across the load is 120 V. Determine the load current if the armature resistance 0.52 Ω.

3.4 A 120-V shunt generator is to maintain constant terminal voltage. At full load the speed drops 10%, as compared to no-load value, and the armature voltage drop is 12 V. The open-circuit characteristic of the generator when driven at 1200 r/min is given as

Generated EMF (V)	52	83	120	134	148
Field current (A)	0.25	0.5	0.75	1.0	1.5

Determine the change in field resistance from no load to full load.

3.5 A short-shunt compound generator delivers 50 A at 500 V to a resistive load. The armature, series field, and shunt field resistances are 0.16, 0.08, and 200 Ω, respectively. Calculate the generated EMF and armature current. If the rotational losses are 520 W, determine the efficiency of the generator.

3.6 A 25-kW 125-V compound generator (short shunt) has a series field in which each coil has 6.5 turns. How many ampere-turns are produced by each coil?

3.7 A short-shunt compound generator has 1000 turns per pole on the shunt field and 4.5 turns per pole on the series field. If the shunt and series field ampere-turns are 1400 and 180, respectively, calculate the power delivered to the load when the terminal voltage is 220 V. Repeat for a long-shunt connection.

3.8 Determine the series number of turns per pole for a 100-kW compound generator to develop 230 V at no load and 230 V at full load. The required ampere-turns per pole are 5800 and 7200, at no load and full load, respectively. Assume a short-shunt connection.

3.9 For the generator in Problem 3.5, calculate the voltage regulation.

3.10 A separately excited generator has a no-load voltage of 148 V when the field current is adjusted to 1.8 A. The speed is 850 r/min. Assume a linear relationship between the field flux and current. Determine:

    a. the generated voltage when the field current is increased to 2.4 A, and

    b. the terminal voltage when the speed is increased to 950 r/min with the field current set at 2.2 A.

4

DC Motors

4.1 Introduction

The dc motor is similar to a dc generator; in fact, the same machine can act as a motor or generator. The only difference is that in a generator the EMF is greater than the terminal voltage, whereas in a motor the generated EMF (counter EMF) is less than the terminal voltage. Thus the power flow is reversed; that is, the motor converts electrical energy into mechanical energy, the reverse process of that of the generator.

Commercial electric power is generated and distributed as alternating-current (ac) power. A remarkably high percentage of this power is ultimately utilized in the form of direct current. Dc motors are highly versatile machines. For example, dc motors are better suited for many industrial processes that demand a high degree of flexibility in the control of speed and torque. In this regard they outperform conventional ac motors operating from a constant-frequency source. The dc motor can provide high starting torques as well as high decelerating torques for applications requiring quick stoppage or reversals.

Speed control over a wide range is relatively easy to achieve in comparison with all other electromechanical energy-conversion devices, in fact, this has traditionally been the dc motor's strength. Although ac drives are replacing many dc applications nowadays, the dc motor is still very important in the automobile industry. Luxury cars, for instance, have well in excess of 50 motors. In this chapter we discuss the dc motor from the point of view of operational characteristics, and its starting and speed control methods.

4.2   Counter EMF in DC Motors

When a voltage is applied to the dc motor, current will flow into the positive brush, through the commutator into the armature winding. The motor armature winding is identical to that of the generator armature winding. Thus conductors under the north field poles carry current in one direction, while all conductors under the south field poles carry the current in the opposite direction. When the armature carries a current it will produce a magnetic field of its own, which will interact with the main field. As we have seen in Chapter 2, the result is that there is a force developed on all conductors, tending to turn the armature.

With the armature rotating as a result of motor action, the armature conductors continually cut through this resultant field. Thus voltages are generated in the very same conductors that experience force action. When operating, the motor is simultaneously acting as a generator. Naturally, motor action is stronger than generator action, because the current is fixed by the polarity of the supply. We illustrate this principle as shown in Figure 4.1 for a two-pole motor. Although the counter EMF opposes the supplied voltage, it cannot exceed the applied voltage. In this respect, the counter EMF serves to limit the current in the armature winding. The armature current will be limited to a value just sufficient to take care of the developed power needed to drive the load.

If no load is connected to the motor shaft, the counter EMF will almost equal the applied voltage. The power developed by the armature in this case is just that power needed to overcome the rotational losses. All this can only mean that the armature current I_A is controlled and limited by the counter EMF E_c ; therefore,

$$I_A = \frac{V_L - E_c}{R_A} \quad \text{A} \tag{4.1}$$

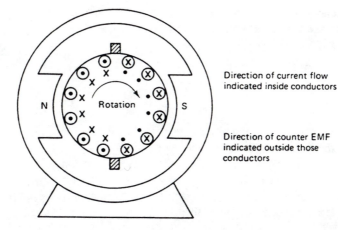

Direction of current flow indicated inside conductors

Direction of counter EMF indicated outside those conductors

FIGURE 4.1.   Armature current direction for clockwise rotation and resulting direction of counter EMF in a dc motor.

where V_L is the line voltage applied across the armature winding, and R_A is the resistance of the armature winding.

Since the counter EMF E_c is an induced or generated voltage, it depends on the flux per pole and the speed of rotation n, in r/min. In other words, $E_c = k_g \phi n$, where the value of k_g is the constant depending on the armature winding and the number of poles on the machine. Substitution of E_c into Eq. 4.1 gives

$$I_A = \frac{V_L - k_g\phi n}{R_A} \tag{4.2}$$

We will refer to this equation later when methods of speed control are discussed.

EXAMPLE 4.1

A dc motor operates at 1680 r/min when drawing 28 A from a 230-V supply. If the armature resistance is 0.25 Ω, calculate the no-load speed if $I_A = 0$ A at no load. (This amounts to assuming that all losses are negligible.)

Solution When loaded, $I_A = 28$ A, therefore,

$$E_c = V_A - I_A R_A = 230 - 28 \times 0.25 = 223 \text{ V}$$

and

$$223 = k_g\phi n = k_g\phi \times 1680$$

or

$$k_g\phi = \frac{223}{1680} = 0.1327$$

At no-load, $I_A = 0$, thus $E_c = V_L = 230$ V. Then $E_c' = k_g \phi n'$, where the prime indicates the new load condition, that is, no load. Hence

$$230 = k_g \phi \times n'$$

and

$$n' = \frac{230}{0.1327} = 1733 \text{ r/min}$$

Note that the flux remained constant. This is true for separately excited and shunt motors if the armature reaction effects are negligible.

Example 4.1 shows that E_c is a large percentage of the applied voltage, which we translate into a high efficiency, as discussed in Section 2.4. Carrying this one step further, if T is the torque (in N·m) exerted on the armature to develop the mechanical power P_d, and n the speed (in r/min), then

$$P_d = \text{mechanical power developed} = \omega T$$

or

$$T = \frac{P_d}{\omega} = \frac{E_c I_A}{\omega} = \frac{k_m \phi \omega I_A}{\omega}$$

Therefore,

$$T = k_m \phi I_A \tag{4.3}$$

This is in agreement with Eq. (2.15).

EXAMPLE 4.2

For the motor of Example 4.1, determine
  a. the developed power under loaded conditions, and
  b. the torque developed under the given load.

Solution
  a.

$$P_d = E_c I_A = 223 \times 28 = 6244 \ \text{W}$$

  b.

$$T = P_d/\omega = 6244/(2\pi \times 1680/60) = 35.5 \ \text{N} \cdot \text{m}$$

4.3   *Classification of DC Motors*

Like the generator there are in general three types of dc motors: series, shunt, and compound. Unlike the series generator, the series motor is widely used because of its excellent starting torque characteristics. Each type of motor has very definite operating characteristics, making it essential to know the load requirements before a proper motor selection can be made. In the separately excited machine, the field winding is connected to a separate supply. If the field is connected in parallel with the armature winding, we obtain a shunt machine. The series machine has a field winding in series with armature winding . As it carries the load current, it is wound with conductor wire of sufficient cross section to enable the winding to carry the load current without excessive heat losses in it.

The compound motor has both field windings. These field windings may be connected so that the fields aid one another, in which case we speak of a cumulative compound motor. Alternatively, they may be connected so that the resulting fluxes oppose, resulting in a differentially compounded motor. A compounded motor may have a long-shunt or a short-shunt connection, depending on whether the shunt field is connected before or after the series field winding. Figure 4.2 shows the three types of connections. As pointed out in Example 4.1, when a dc motor is loaded it tends to slow down. This we may know from practical experience. For example, a drill will slow down once the drill bit starts to cut steel. The amount by which a motor slows down at full load as compared to the no-load condition depends greatly on the type of field winding connection employed.

For the shunt motor, for example, the speed may drop by approximately 8% when the mechanical load is increased to full load. For all practical purposes, this may be considered

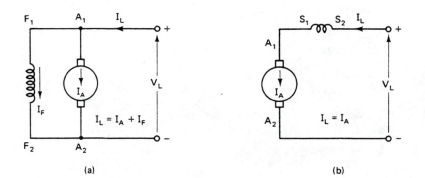

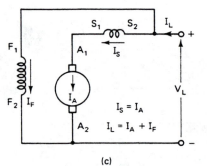

FIGURE 4.2. **Classification of dc motors (terminal markings follow accepted practices): (a) shunt motor; (b) series motor; (c) compound motor, long shunt.**

a constant-speed type of motor. Naturally, these figures are somewhat arbitrary; they are used to differentiate motors based on speed variation. Series and compound motors with strong series fields are generally considered variable speed motors; their speeds change considerably more.

Of course, it is possible to change the motor speed by a controller of some kind. This normally involves an adjustment either by an operator or by electronic means, depending on the process. In such instances we refer to speed control. More on this subject will be said in later sections.

4.4   *Starting DC Motors*

At the instant of startup, the counter EMF E_c is zero because the armature is not rotating. If we start the motor of Example 4.1 directly across the 230-V supply line, the armature current would tend to reach $230/0.25 = 920$ A. Such a current, besides subjecting the armature to a severe mechanical shock, would blow fuses, thereby disconnecting itself from the supply. It is therefore necessary to insert some resistance in series with the armature circuit to limit the current. As the motor comes up to speed, this resistance is taken out in steps because E_c rises as the motor comes up to full speed. A practical arrangement that accomplishes this is called a *starter*. If the starting current in the example above is to be limited to, say, 60 A, the total resistance of the starter plus the armature winding must be $230/60 = 3.83 \; \Omega$.

Figure 4.3 illustrates in principle one of a variety of manual starters available for a shunt-connected dc motor. The starter must be rated on the basis of the horsepower rating of the motor and the voltage at which it is used. For small motors (less than 1 hp) no starter is generally required. Such motors have a high enough armature resistance and inductance to limit the current. In addition, the mass of the rotational members is usually small, so that they quickly accelerate.

To start the motor in Figure 4.3, the main switch is closed and the starter arm is moved to contact 1. After the armature has accelerated sufficiently on the first contact, the starter arm is slowly moved to the following contacts, until the iron keeper on the arm is held by

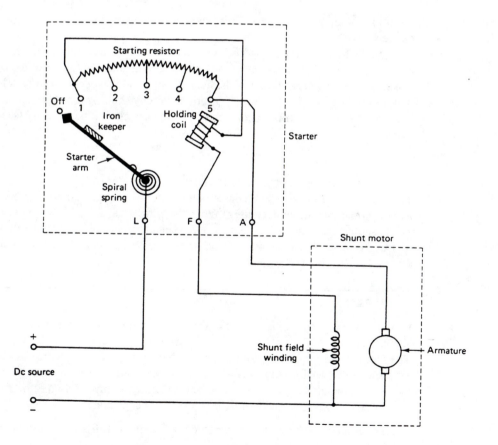

FIGURE 4.3. Manual starter for shunt motor.

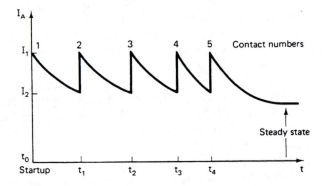

FIGURE 4.4.   Variation in armature current during startup period.

the holding-coil electromagnet. This whole process should take a few seconds, depending on the size of the motor.

In the final position, no added resistance is in the armature circuit. The starter arm is kept in the last contact position by the holding coil against the force exerted by the spring. When there is a power failure or if the field circuit should be opened accidentally, the spring will return the starter arm to the off position, thereby shutting down the motor.

With the starter arm on contact 1 the armature current becomes I_1 (see Figure 4.4), where

$$I_1 = \frac{V_{supply}}{R_A + R_{starter}} \tag{4.4}$$

Since the torque is proportional to the current and flux, it follows that maximum torque is available at startup to accelerate the armature. During the acceleration period, E_c increases and the armature current decreases. When the current has fallen to some preselected value I_2, the starter arm is moved to contact 2. Sufficient resistance is cut out to allow the current to rise to I_1 again. This operation is continued until the last contact is reached and the motor assumes its steady-state speed and current.

EXAMPLE 4.3

Calculate the required resistances for a four-step starter to limit the starting current of a dc shunt motor to 150% of rated current. Assume all four steps have equal resistance values. The motor is 25 hp, 240 V, 860 r/min and has an armature resistance of 0.08 Ω and an efficiency of 89%. Determine at which speeds the starter resistors must be cut out so that the current remains at or above rated value during startup. Assume the field current is negligible compared to rated armature current and constant.

Solution   Figure 4.3 shows the motor starting circuit. At full load the motor input power

$$P_{in} = \frac{hp \times 746}{\eta} = \frac{25 \times 746}{0.89} = 20,955 \text{ W}$$

and the line current

$$I_L = \frac{20,955}{240} = 87.3 \text{ A}$$

This current will be limited to 150%, or

$$I_{max} = 1.5 \times 87.3 = 131 \text{ A}$$

The current-limiting resistance at startup is then

$$I_{max} = \frac{V_{supply}}{R_A + 4R} = \frac{240}{0.08 + 4R} = 131 \text{ A}$$

and $R = 0.438\Omega$. Furthermore, at rated speed and current,

$$E_c = V - I_A R_A = k_g \phi n$$

or

$$E_c = 240 - 87.3 \times 0.08 = k\phi \times 860$$

Therefore, $k_g \phi = 0.271$, which is constant under our stated assumptions. At startup, with the starter arm at contact 1, $I_{max} = I_R$ and there are four resistors in series with the armature. Then, $240 = I_{max}(0.08 + 4R)$, since $n = 0$ at startup.

The motor will accelerate and the current decay until $I_A = I_{Arated}$, at which point the starter arm is moved to contact 2. At this location there are three resistors in series with the armature and $I_A = I_{A\,max}$. Then

$$240 = 131(0.08 + 3 \times 0.438) + 0.271n$$

from which

$$n = 212 \text{ r/min} \quad \text{and} \quad E_c = 57.4 \text{ V} \text{ at time } t_1$$

Similarly, on contact 3, with $2R$ in the circuit,

$$n = 423 \text{ r/min} \quad \text{and} \quad E_c = 114.8 \text{ V} \text{ at time } t_2$$

On contact 4, with R in the circuit,

$$n = 635 \text{ r/min} \quad \text{and} \quad E_c = 172.1 \text{ V} \text{ at time } t_3$$

On contact 5, the final starter arm position with no additional resistance in the armature circuit,

$$n = 847 \text{ r/min} \quad \text{and} \quad E_c = 229.5 \text{ V} \text{ at time } t_4$$

after which the motor will reach steady-state condition. The time intervals between contact positions will depend on motor plus load inertia. The variation of armature current versus time will be somewhat similar to that depicted in Figure 4.4.

Automatic Starting of DC Motors

Apparent disadvantages of the manual starters are the need for manual operation, bulkiness of the starter, lack of remote starting operation, and possibility of improper operation, that is, nonuniform acceleration. Automatic starters overcome all these disadvantages in addition to providing other desirable control features, such as undervoltage protection.

To illustrate these points, let us refer to a simple automatic starter connected to a shunt motor as shown in Figure 4.5. This starter circuit is called a *counter-EMF starter*, in which for simplicity the starting resistance is cut out in a single step. Contactor M is a magnetic contactor. It has a coil placed on an iron core so that when current flows through the coil, the iron becomes magnetized. This attracts a movable iron armature that carries a set of contacts electrically insulated from each other. As the armature moves towards the core, the moving contacts are moved towards stationary contacts, thereby closing the circuit to which they are connected in series. In Figure 4.5 the contactor is energized by pressing the start button, which together with the stop button may be remotely located from the motor. With main contactor M energized, contacts M and M_1 close. Contact M_1, called a *holding contact*, keeps contactor M energized after the start button is released. The motor is then connected to the supply lines with the starter resistance R in series with the armature. Because the counter EMF is zero at starting, the IR drop across the armature is too low to energize contactor A, called the *accelerating contactor.*

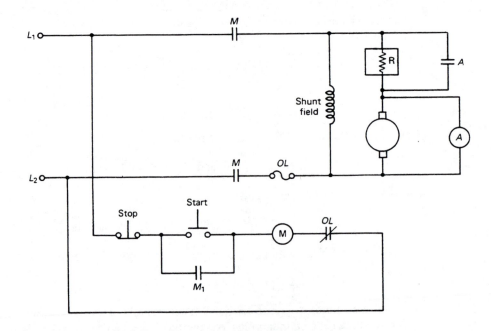

FIGURE 4.5.   Wiring diagram of a counter-EMF starter connected to a shunt motor.

As the motor speeds up, the counter EMF increases, and the voltage across the armature increases. When E_c reaches about 80% of line voltage, contactor A closes and its associated contact A shorts out the starter resistor R. Now the motor is directly across the line. To obtain smoother acceleration and closer control of the armature current, several contactors may be arranged to close on different voltages. In this way the starting resistance is cut out in several steps.

If the motor load is different at each startup, the counter EMF will not build up in a similar way each time. This may cause closure of the accelerating contactor A too soon or too late. To overcome this problem, definite-time starters may be used. These are starters in which the cutout steps are controlled by time-delay relays. In the circuit in Figure 4.5, overload protection is provided by the thermal overload sensor and relay, OL. There are basically two types; both operate from the heat generated in a heating element through which the motor current flows. One type uses a bimetallic strip; the other melts a strip of solder. Both act to open the motor control circuit. To stop the motor under normal conditions, the stop button is pressed, thereby opening the control circuit, which causes contactor M to drop out. The motor is then disconnected from the supply.

Undervoltage protection is provided by contactor M. If the line voltage should drop below a certain value or in the event of power failure, the contactor drops out, thereby disconnecting the motor from the line. Restarting the motor would require pressing the start button.

4.5 *Speed Characteristics of DC Motors*

When the mechanical load is removed from the motor, the motor speed will increase. The amount by which it increases depends on the type of motor. The speed of a shunt motor increases about 8%; for a compound (cumulative) motor it rises appproximately 15 to 20%. For the series motor it would rise very rapidly, and it is for this reason that series motors must always drive some load. In order to understand why this happens, let us rewrite Eq. 4.2 in the form,

$$n = \frac{V - I_A R_A}{k_g \phi} \tag{4.5}$$

In shunt motors the flux ϕ is only slightly affected by the armature current, while the $I_A R_A$ drop rarely exceeds 5% of line voltage. Therefore, the maximum change in speed must be of the same order as I_A for a shunt motor. Hence, shunt motors are fairly constant-speed machines, as indicated by curve a in Figure 4.6.

For a compound motor, when the load is removed, two factors affect the speed, namely I_A and ϕ. Unlike the shunt motor, the effect of the series field is removed under no-load conditions, thereby weakening the overall field flux. The result is then a larger increase in speed, since the speed is inversely proportional to the flux, as Eq. 4.5 indicates. This is represented by curve b in Figure 4.6. We can now see why a series motor will run at dangerously high speeds when the load is removed. There would not be any flux apart from the residual magnetism, because the flux depends on the load current. Curve c in Figure 4.6 illustrates the speed versus load current behavior for the series motor. It is for this reason

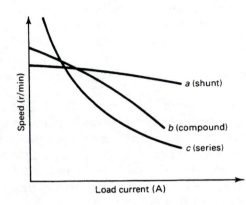

FIGURE 4.6.   Comparative speed-load characteristics of shunt, compound, and series motors.

that series motors are not used in instances where the load can be accidentally disconnected. For example, belt-coupled drives should never be used with series motors.

In a similar fashion, the torque-load characteristic for the shunt, compound, and series motor can be compared. Figure 4.7 shows this for the three dc motors. Since $T = k_m \phi I_A$, the torque of a shunt motor depends only on the armature current. The graph is thus a straight line. For the compound motor (cumulative), the series field increases with loading, aiding the shunt field. The resulting graph is somewhat curved, as shown. In a series motor, neglecting saturation, the developed torque depends on the load current squared, because the flux in turn depends on the current. Under this condition,

$$T = k_m \phi I_A = k_m (k_1 I_A) I_A = k' I_A^2 \tag{4.6}$$

which is a parabolic equation. If magnetic saturation is considered, the graph will become a straight line at heavy loads, since the flux will not change with increased loading. From Figure 4.7 it is evident that for a given current below full-load value, the shunt motor develops the largest torque. For currents exceeding rated current, the series motor develops the largest torque. It is for this reason that in applications requiring large starting current, such as for hoists, electric trains, cranes, and so on, the series motor is the most suitable machine.

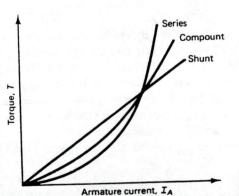

FIGURE 4.7.   Comparative torque-load characteristics for the three dc motors.

EXAMPLE 4.4

A 240-V shunt motor has an armature resistance of 0.25 Ω. Under load, the armature current is 24 A. Suppose the flux is suddenly decreased by 2.5%; what would be the immediate effect on the developed torque?

Solution   When the current is 24 A,

$$T = k_m \phi I_A = k_m \phi \times 24$$

and

$$E_c = V_L - I_A R_A = 240 - 24 \times 0.25 = 234 \text{ V}$$

If ϕ is suddenly decreased by 2.5%, E_c is also decreased, since $E_c = k_g \phi n$ and the speed cannot change instantaneously. Thus,

$$E_c = 0.975 \times 234 = 228.15 \text{ V}$$

The new armature current is

$$I_A' = \frac{V_L - E_c}{R_A} = \frac{240 - 228.15}{0.25} = 47.4 \text{ A}$$

The new value of developed torque is then

$$T' = k_m(0.975\phi) \times 47.4 = 46.215 k_m \phi$$

This corresponds to an increase in torque of

$$\frac{T'}{T} = \frac{46.215 k_m \phi}{24 k_m \phi} = 1.93 \text{ times}$$

Thus a slight decrease of flux almost doubles the torque. This increased torque causes the armature to accelerate to a higher speed. At that point the counter EMF limits the armature current just enough to carry the load at this new speed.

EXAMPLE 4.5

For Example 4.4, determine the new steady-state speed after the field has been decreased. Assume the motor was operating at 640 r/min before the field was adjusted.

Solution   After steady-state is reached, the torque will have decreased to assume its original value, namely,

$$T = k_m \phi \times 24$$

Therefore, in the new steady-state condition, the flux is

$$\phi' = 0.975 \times \phi$$

and

$$T = k_m \phi' I'_A = k_m (0.975\phi) I'_A$$

or the new armature current is

$$I'_A = \frac{24}{0.975} = 24.62 \ \text{A}$$

and

$$E'_c = 240 - 24.62 \times 0.25 = 233.8 \ \text{V} = k_g \phi' n'$$

Hence

$$\frac{233.8}{234} = \frac{0.975 k_g \phi n'}{k_g \phi \times 640}$$

from which the new steady-state speed is

$$n' = 656 \ \text{r/min}$$

Speed Regulation

The classification of motors on the basis of how the speed changes with load is particularly important in the selection of a motor to do a specific task. In applications such as grinders and polishers, it has been found that constant speed shunt motors are the best.

Variable-speed compound motors, on the other hand, are most satisfactory when driving loads requiring considerable torque at startup or when subjected to impulse loads. Examples of these are compressors and pumps. Series motors, on the other hand, are usually employed where starting torques and acceleration requirements are severe. Mine hoists and street cars fit this category. If the speed of a motor is relatively constant over its normal range, we speak of the motor as having good speed regulation. Speed regulation is usually expressed as a percentage and is found as follows:

$$\begin{aligned} \text{speed regulation} &= \frac{\text{no-load speed} - \text{full-load speed}}{\text{full-load speed}} \times 100\% \\ &= \frac{n_{NL} - n_{FL}}{n_{FL}} \times 100\% \end{aligned} \qquad (4.7)$$

where n_{NL} is the no-load speed and n_{FL} is the full-load speed, both expressed in r/min. It should be understood that speed changes are the result of changes in loading conditions, not those as the result of speed adjustments made manually or otherwise by operating personnel.

EXAMPLE 4.6

In Example 4.1 the speed at full load was 1680 r/min, while the calculated no-load speed was 1733 r/min. Determine the speed regulation for this motor.

Solution   The speed regulation is

$$\frac{n_{NL} - n_{FL}}{n_{FL}} \times 100 = \frac{1733 - 1680}{1680} \times 100 = 3.2\%$$

4.6   *Speed Control of DC Motors*

Equation 4.5 gave us the expression for the speed, namely,

$$n = \frac{V - I_A R_A}{k_g \phi} \tag{4.8}$$

In order to change the speed of a dc motor, we see that there are three quantities which may be considered as parameters: the armature resistance R_A, the flux ϕ, and the terminal voltage V.

Adding a resistor in series with the armature effectively increases the armature circuit resistance. Equation 4.8 indicates that this results in a reduction of the steady-state speed, since the numerator becomes smaller. The no-load speed is not affected, since $I_A = 0$ then. In this method of control the field current is kept constant. Figure 4.8 illustrates a family of characteristics for different values of armature circuit resistances. As is apparent, this method of speed control is relatively simple and inexpensive. It has some disadvantages, however, as can be seen from Figure 4.8.

1. By adding resistance in the armature circuit, the speed of the motor, compared to that without connecting any resistance, is always lower.
2. This method of speed control is ineffective at no load.
3. Adding a resistance means increased $I^2 R$ losses, therefore wasted power. As a rule of thumb, the percentage reduction in speed equals the percentage of power input that is consumed in the added resistor.

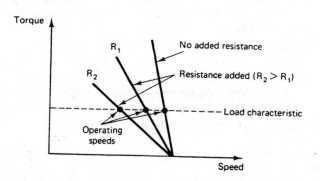

FIGURE 4.8.   Armature resistance speed control of dc motors.

4. The constant-speed characteristic of the motor is lost. Speed control by this method is generally limited to 50% of rated speed.

EXAMPLE 4.7

A 240-V shunt motor runs at 850 r/min when the armature current is 70 A. The armature circuit resistance is 0.10 Ω. Calculate the required resistance to be placed in series with the armature to reduce the speed to 650 r/min when the armature current is then 50 A.

Solution The initial counter EMF is

$$E_c = 240 - 70 \times 0.1 = 233 \ \text{V}$$

Since the field current remains constant, the counter EMF is proportional to the speed. Therefore, the counter EMF at 650 r/min is

$$E'_c = 233 \times \frac{650}{850} = 178.2 \ \text{V}$$

Thus the voltage drop in the armature circuit amounts to

$$V_{R+R_A} = 240 - 178.2 = 61.8 \ \text{V}$$

The total armature circuit resistance is then

$$R + R_A = \frac{61.8}{50} = 1.24 \ \Omega$$

Therefore, the additional resistance required in the armature circuit is

$$R = 1.24 - 0.1 = 1.14 \ \Omega$$

The second method of speed control is by changing the flux ϕ. To do so we connect a resistance in series with the field winding. Normally, this is a variable resistor, and it is generally called a *field rheostat*. As can be appreciated, we can add resistance only; therefore, the field current is decreased and Eq. 4.8 tells us that the speed increases with a reduction in ϕ. Thus this method, too, has some disadvantages. One is that we can only raise the speed at which the motor normally runs at a particular load. This is shown in Figure 4.9. Another disadvantage is that the speed is increased without a corresponding reduction in shaft load, so we will be overloading the motor. However, at light loads or no-load conditions, the speed can be varied above normal operating speeds by about 300%.

EXAMPLE 4.8

A 240-V dc shunt motor runs at 800 r/min at no load. Determine the resistance to be placed in series with the field so that the motor runs at 950 r/min when taking an armature current

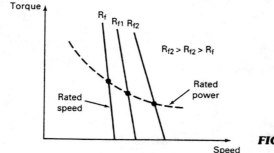

FIGURE 4.9.   Field current control.

of 20 A. Field resistance is 160 Ω, and the armature resistance 0.4 Ω. Assume the pole flux is proportional to the field current and neglect rotational losses.

Solution   At no load, since the rotational losses are assumed negligible,

$$E_c = 240 \text{ V} \quad (I_A = 0) \quad \text{and} \quad k_g\phi = \frac{240}{800}$$

The field current

$$I_F = \frac{240}{160} = 1.5 \text{ A}$$

With added field resistance,

$$E_c = 240 - 20 \times 0.4 = 232 \text{ V} = k_g\phi' \times 950$$

With decreasing field flux (added resistance R in the field circuit reduces the field current), the speed increases in accordance with Eq. 4.8. Therefore,

$$232 = k_g\phi \times \frac{I'_F}{1.5} \times 950 \quad I'_F = 1.221 \text{ A}$$

Hence

$$R + 160 = \frac{240}{1.221} = 197 \text{ } \Omega$$

and the added resistance

$$R = 37 \text{ } \Omega$$

The third method of speed control is by changing the terminal voltage V of the motor. This is normally the most frequent application of control, at least for shunt motors where the field winding is then separately excited. The results according to Eq. 4.8 are represented in Figure 4.10. The voltage control method lowers the speed in a similar fashion as the armature circuit resistance speed control method; however, it does not have its drawbacks. The no-load speed and full-load speed can be reduced all the way down to zero, if desired.

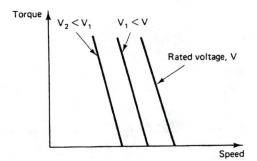

FIGURE 4.10. **Armature voltage speed control.**

Not much additional power is wasted, as the virtual constant slope of the characteristic curves attest.

There are various ways to obtain a variable dc voltage. Some modern ways using solid-state devices are deferred to later sections. Finally, if speed control above and below normal speeds are desired, we must resort to a combination of two methods. For example, field control combined with armature voltage control would achieve this mode of control.

4.7 Applications of DC Motors

Dc motors are widely used in practice, particularly in applications where accurate control of speed or position of the load is required. They are used in conveyor drives, machine tools, packaging and cutting machines, and so on. At one time the use of the dc motor hinged on the availability of a dc supply. However, modern solid-state control circuits use diodes and thyristors and have overcome these drawbacks. In fact, often it is more economical and easier to obtain dc from a rectified ac source.

Choosing a motor for a particular application may be straightforward or more involved, depending on the factors to be considered. If speed ranges over a wide speed range, say up to 6:1, are required, the obvious choice at one time would have been the dc motor. Control over this range of speed requires a sophisticated controller involving both field and armature control. At present, variable-speed, variable-frequency controllers are available for ac motors which would be cost competitive with dc motors.

It is interesting to note that the maximum torque in a dc motor is limited by commutation and not, as with other types of motors, by heating. If the armature current of a dc motor is allowed to become excessive, sparking at the brushes can become so severe as to cause flashover between brushes. This renders the motor useless. Another point of interest, as expected, is the higher starting torque of the compound motor in comparison with the shunt motor. This feature is attributable to the series field winding.

In any event, the type of motor selected is usually dictated by the type of load characteristic. For instance, a subway car requires high starting torque but relatively little running torque, so that the series motor is desirable. This application is normally referred to as a *traction drive*. It should be mentioned that electric trains do not have a single motor, but often each wheel or pair of wheels is driven by a separate motor. This prevents the size of the motor from becoming prohibitive and avoids slippage at the wheels.

A crane motor might require a high starting torque and running torque to lift a load, then a shunt motor would be a correct choice. Also, fan drives would be a suitable choice for shunt motors. At startup they do not require much torque, but as the speed picks up, the torque requirement varies with the square of the speed.

Furthermore, the rating of a motor depends on the conditions under which it is used. Often it is necessary to run a motor beyond its rated value for extended periods, depending on the size motor. Because of its large thermal mass, it will take a considerable amount of excess current to cause sufficient heating to burn it out. Small motors, however, may burn out in a few minutes on severe overload. The best advice in choosing a motor subjected to these conditions would be to consult the manufacturer.

4.8 Solid-State Controllers

We have seen that the torque-speed characteristic of a dc motor can be controlled by adjusting the armature voltage, by adjusting the field current, and by inserting resistance into the armature circuit.

Today's trend is toward solid-state control. A variable dc voltage can be obtained from a fixed dc voltage source by means of a dc chopper. When the source is ac power, the variable dc voltage is obtained by means of a converter or controlled rectifier. The efficiencies encountered in such power electronic circuits are very high, normally in the high 90% range. In practice, two controllers used often are the armature voltage controls (for speeds below base speed) and field voltage controls (for speeds above base speed). With these controllers it is also possible to reverse the power flow through the controller, so that regenerative braking can be achieved.

It is our purpose in this section to introduce some of the concepts without going in too much detail. It will be apparent that detailed application of power electronic controls to electrical machines requires a sound knowledge of both disciplines. The accelerating use of such drive systems makes a basic understanding of this type of control increasingly important.

Choppers

A special class of solid-state circuits has been developed for controlling dc motors which are supplied from fixed dc voltage sources. They are used in applications where the source is a battery; for instance, in industrial electrically powered trucks. The circuits, called *choppers*, are used to replace switched series armature-circuit resistors. The advantage is higher efficiency and continuous control.

The semiconductor devices used in choppers may be power transistors, forced-commutated thyristors, or gate turn-off thyristors (GTOs). In essence, they control the average

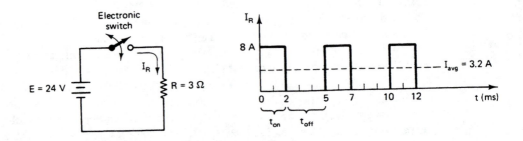

FIGURE 4.11.   **Current waveform through R with switch closed for 2 ms and opened 3 ms, repetitively.**

voltage supplied to the motor. The combination of control system and motor is usually referred to as the *drive system*. Basically, the chopper acts as a switch that is turned on and off in regular cycles. Figure 4.11 shows a chopper connected to a resistive load. If a voltmeter is connected across the load resistor R, the average voltage read by a dc voltmeter would be

$$V_R = \frac{t_{on}}{t_{on} + t_{off}} \times E = \delta \times E \quad \text{V} \qquad (4.9)$$

where δ is the ratio $t_{on}/(t_{on} + t_{off})$ and is called the *duty cycle* of the waveform, and by controlling it, the average voltage to the load resistance R is controlled. For example, in Figure 4.11, the on-time is 2 ms, while the off-time is 3 ms. Thus the average current in the circuit is

$$I_R = \frac{E}{R} \times \frac{t_{on}}{t_{off} + t_{on}} = \frac{24}{3} \times \frac{2}{2+3} = 3.2 \quad \text{A}$$

For the specific switching rate, "on" 2 ms and "off" 3 ms, $1/(5 \times 10^{-3}) = 200$ on-off operations per second are required. It can readily be appreciated that at these switching rates or greater, electronic switching techniques must be used. For this, power electronic components are used to handle the required power. There are numerous varieties of circuits that accomplish these tasks; however, we must resort to the principles involved and therefore examine only some basic circuits.

Before examining some of these methods, the method of speed control by controlling the duty cycle will be illustrated in the following example. A simplifying assumption will be made, namely, that the necessary filtering is included or that the armature inductance is large enough to provide this function. This implies that the resulting armature current I_A is practically constant at the average value.

EXAMPLE 4.9

A dc motor is supplied from a 120-V dc switched source as shown in Figure 4.12. The duty cycle is 50%, and the input power is 5 kW at 600 r/min. The armature circuit resistance is 0.048Ω. Determine

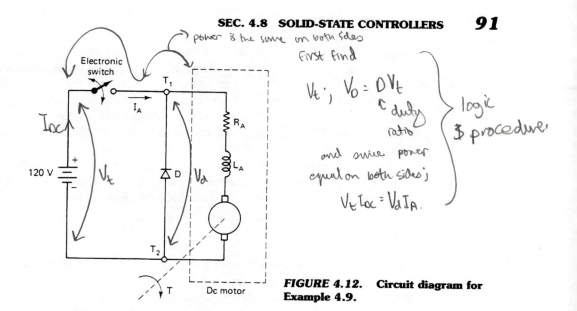

FIGURE 4.12. **Circuit diagram for Example 4.9.**

Handwritten annotations:
> power is the same on both sides
> First find
> $$V_t; \ V_0 = D V_t$$
> c duty ratio
> and since power equal on both sides;
> $$V_t I_{DC} = V_d I_A.$$
> logic & procedure.

a. the delivered shaft horsepower, and

b. the new speed and output horsepower if the duty cycle is increased to 0.6.

Solution

a. If the duty cycle is 0.5, the average motor voltage is $V_M = 0.5 \times 120 = 60$ V and the average input current is $5000/120 = 41.7$ A. Since the average power is the same on either side of the switch, the average motor current is $120 \times I_{DC} = 60 \times I_A$, from which

$$I_A = 2 I_{DC} = 2 \times 41.7 = 83.4 \text{ A}$$

The average counter EMF is

$$E_c = 60 - 83.4 \times 0.048 = 56.0 \text{ V} = k_g \phi n$$

Hence $k_g \phi = 56/600$. Therefore, the average output power is

$$56.0 \times 83.4 = 4670 \text{ W} \quad (6.26 \text{ hp})$$

b. When the duty cycle is now adjusted to 0.6, the average motor voltage becomes $0.6 \times 120 = 72$ V. With the load remaining the same, the torque will remain constant since $T = k_m \phi I_A$, and the field flux is constant. Thus I_A remains the same. Hence

$$E_c = 72 - 0.048 \times 83.4 = 68 \text{ V} = k_g \phi n'$$

and the new speed

$$n' = \frac{68}{56} \times 600 = 729 \text{ r/min}$$

The output power

$$P_o = E_c I_A = 68 \times 83.4 = 5671 \text{ W} \quad (7.6 \text{ hp})$$

Example 4.9 illustrates the means of controlling the speed of the motor by controlling the duty cycle rather than by changing the armature resistance or field resistance. Control by electronic means results in insignificant added losses; therefore, high system efficiencies can be maintained at all speeds. Considering that at a duty cycle of 1.0 (that is, when full voltage is applied), the speed becomes 1243 r/min, the speed of the motor can be controlled over a wide range. Of course, additional control is available by changing the field current as well.

When a high inductive circuit is interrupted abruptly, the inductance attempts to keep the current flowing. It can do so only when a path is provided. Therefore, a diode normally referred to as a *free-wheeling diode*, is added to the circuit as shown.

Diodes

A *silicon diode* is a two-terminal device consisting of a thin layer of silicon, doped so as to provide a P-type and N-type layer. A diode allows unidirectional flow of current; that is, it conducts well in only one direction. Although the actual volt-ampere characteristic of a diode is nonlinear and depends on the temperature, for many applications it can be represented as a switch that conducts current when forward-biased and blocks currents when reversed-biased. Its circuit model and electrical characteristic is shown in Figure 4.13. Current flows through the diode when the voltage at the anode (P-type material) is positive with respect to the cathode (N-type material). The arrow configuration of the symbol shows the direction of current flow. If a voltage of opposite polarity to that shown is applied to the diode, a negligible current will flow (for all practical purposes), and the diode is than open circuited.

Half-Wave Diode Rectifiers

The simplest rectifier circuit consists of a single diode operating into a resistive load, as shown in Figure 4.14. The input voltage v_{in} is sinusoidal. During the positive half cycle

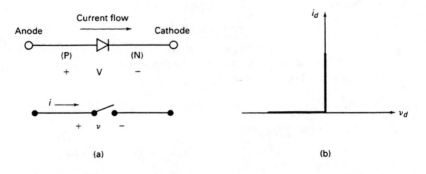

FIGURE 4.13.   (a) Diode symbol and model; (b) ideal characteristic.

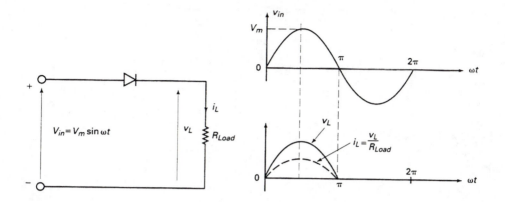

FIGURE 4.14.   Half-wave diode rectifier.

of the input voltage, the diode acts like a closed switch, according to the diode model of Figure 4.13, and conducts the load current i_L. This happens during the time interval from $\omega t = 0$ to π. The load voltage v_L during this period equals the input voltage v_{in}, because the diode forward voltage drop is assumed to be zero here. When the applied voltage is negative, during $\omega t = \pi$ to 2π, the diode acts like an open switch and therefore conducts zero current, and the load voltage is zero.

Thyristors

An important member of the thyristor family is the silicon-controlled rectifier (SCR). The thyristor is a three-terminal device made from four layers of alternating P- and N-type materials. The terminals are called *anode, cathode,* and *gate* (see Figure 4.15a). The device operation is similar to a diode except that it must be turned on by a signal applied to its gate lead before it will begin conduction of current when it is forward-biased. The circuit model is a switch as shown in Figure 4.15b, and the characteristic for the model is shown in Figure 4.15c. The switch is open for either direction of current flow before a gate signal is applied. When forward voltage and a gate signal are applied, the switch is closed and the thyristor model carries forward current with no voltage drop. Gate pulses are of a few ms duration and of relatively low power compared to that controlled. Once started, the gate signal can be removed as it loses control. The current cannot be switched off by another pulse. The thyristor remains in conduction until the anode-cathode current i falls below a level called the *holding current*. In ac-powered circuits this happens by the natural reversal of the ac supply voltage, which reduces the thyristor current to zero, and then applying the reverse bias. The current remains zero and the thyristor is effectively turned off. This process of interrupting current flow and restoring the thyristor to its nonconducting, or blocking, or off-state, is known as *natural cummutation* or *line commutation*. Natural commutated thyristors are normally used in ac-dc converters.

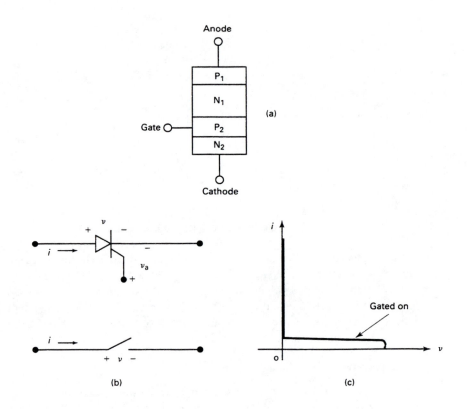

FIGURE 4.15. (a) Thyristor construction; (b) symbol; (c) characteristic.

Controlled Rectifiers

The application of diodes for the conversion of ac power to dc power is known as *rectification*. Circuits using diodes provide a constant dc output voltage, whereas circuits using controlled switching devices provide variable output voltages, and as such are known as *controlled rectifiers*. Therefore, replacing the diode in Figure 4.14 by a thyristor (SCR) results in a controlled rectifier (see Figure 4.16a). The circuitry for applying the gating pulses to the thyristor is not shown in this figure. If the thyristor is triggered (that is, the gate current applied) at $\omega t = \alpha$, it will conduct for the remainder of the positive half-cycle of the supply voltage. The gate pulse is applied at the so-called firing angle α, which is measured from the zero crossing of the supply voltage. The thyristor thus conducts during the time interval α to π, called the *conduction interval*. Once fired, the load voltage equals the supply voltage, $v_L = v_{in}$.

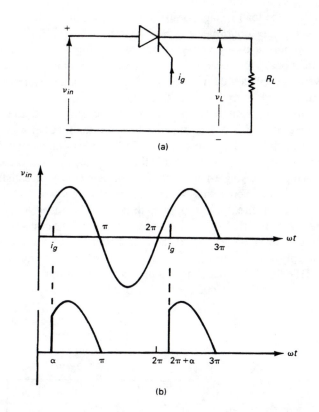

FIGURE 4.16.   Controlled half-wave rectifier: (a) circuit; (b) waveforms.

Since the input voltage is ac, the thyristor current goes through a natural zero, and a reverse voltage will appear across the thyristor. The device is then turned off. The firing angle α can be changed from zero to π, which controls the output voltage, since the average value of the load voltage V_L can be shown to be

$$V_L = \frac{V_m}{2\pi}(1 + \cos\alpha) \tag{4.10}$$

Note that when the thyristor is triggered at $\alpha = 0$, the average load voltage is $V_L = V_m/\pi$, then the output voltage is the same as that of the half-wave diode rectifier, since the thyristor then behaves like a diode.

Three-Phase Rectifiers

The thyristor or SCR is really an electronic switch that can operate over part of an ac cycle. By using this switching capability synchronized to the ac line, it is possible to vary accurately the load voltage, or in the event the load is a dc motor, its current, torque, speed, or power can be controlled.

As the resulting voltage waveform in Figure 4.16b indicates, the current flow is unidirectional. However, since no filtering is included, the ripple content is large. If the converter supplies a dc motor, this in turn causes the hysteresis and eddy currents in the magnetic material of the motor to be high; consequently, the losses are increased.

To improve motor efficiency and reduce harmonics, an inductor is often used as part of the rectifier. It is placed in series between the rectifier and motor. Since it has a small resistance, the voltage drop across the inductor is minimal, but the ac component, or ripple, is greatly suppressed. The motor is therefore able to run more efficiently, hence cooler.

Full-wave, single-phase rectifiers or converters are satisfactory for small motors up to about 5 hp. Above this rating, line currents become excessive and motors should be operated from three-phase supplies to provide balanced loading on the network. Figure 4.17a illustrates a half-wave three-phase rectifier. The circuits show three diodes connected to a common resistive load R; the load current is returned to the neutral n of the supply. The supply voltages are v_{an}, v_{bn}, and v_{cn}, as shown in Figure 4.17b.

The diodes conduct one at a time for 120° each, in sequence. At any given instant in time, the diode that conducts is the one that is connected to the highest instantaneous supply

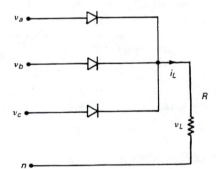

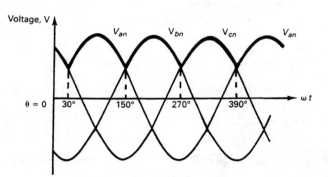

FIGURE 4.17. Three-phase half-wave rectifier: (a) circuit; (b) waveforms.

voltage. Since the common terminal of the cathodes is thereby raised to the same voltage, it will reverse-bias the remaining two diodes. As Figure 4.17b shows, the load voltage follows the envelope of the highest instantaneous supply voltage. There are numerous three-phase rectifier circuits using both diodes and thyristors. In diode circuits, the load voltage is proportional to the ac source voltage, while in thyristor circuits the load voltage is controllable.

As can be seen, three-phase rectifier circuits provide more voltage pulses per cycle of line frequency. This implies that the current flows over a larger portion of the cycle, increasing the average-to-rms current, thereby reducing the heating effects in case of a motor load.

DC Motor Speed Control

As an example of a simple dc motor speed control problem using power electronic components refer to Figure 4.18, which shows a half-wave SCR motor control. The gate-signalling characteristic is used for speed selection (by the setting of the potentiometer) and to provide

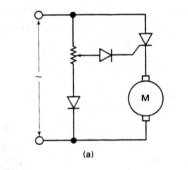

(a)

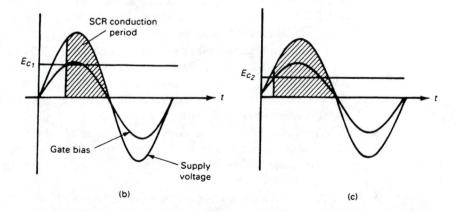

(b)

(c)

FIGURE 4.18. **(a) SCR motor control; (b) motor lightly loaded; (c) motor load increased from that in part (b).**

feedback to compensate for changes in speed with varying loads. This particular circuit uses the counter EMF of the motor as a feedback voltage ($E_c \propto n$). The gate firing occurs when the potentiometer wiper voltage exceeds the counter EMF developed by the motor. At that instant and for the remaining part of the half cycle, the applied voltage is fed to the motor. If the motor load increases, the motor slows down and the counter EMF is lower. This causes the SCR to fire sooner in the cycle and therefore will be on longer. Figure 4.18b and c show this principle and illustrate the increased power delivered by the source to the motor, by means of the increase in the shaded areas.

PROBLEMS

4.1 A 240-V dc motor takes an armature current of 20 A when running at 960 r/min. The armature resistance is 0.2 Ω. Determine the no-load speed, assuming negligible losses.

4.2 What would be the no-load speed in Problem 4.1 if the armature current is 1 A at no load?

4.3 A 120-V shunt motor has the following parameters: $R_A = 0.40\ \Omega$, $R_F = 120\Omega$, and rotational loss 240 W. On full load the line current is 19.5 A and the motor runs at 1200 r/min. Determine

    a. the developed power,
    b. the output power,
    c. the output toque, and
    d. the efficiency at full load

4.4 Calculate the required resistances for a three-step starter to limit the starting current to 140% rated current. Assume the motor data from those of Example 4.3. Determine the motor speed at each step and the final steady-state speed.

4.5 A 460-V dc motor drives a 50-hp load at 900 r/min. The shunt field resistance is 230 Ω and the armature resistance is 0.24 Ω. If the motor efficiency is 88%, determine (a) the rotational losses, and (b) the speed regulation.

4.6 A dc shunt machine generates 230 V on open circuit at 1200 r/min. The armature resistance is 0.5 Ω, and the field resistance is 115 Ω. When run as a motor on no load it takes 3.6 A at 230 V. Calculate the speed and efficiency of the machine run as a motor when the line current is 50 A. Assume at this load the armature reaction effect weakens the field by 6%.

4.7 A series motor takes 40 A at 460 V while hoisting a load at 6 m/s. The armature plus field resistance is 0.48 Ω. Determine the resistance to be placed in series with the motor to slow the hoisting speed to 4 m/s. Assume linear operation on the magnetization curve.

4.8 A 120-V separately excited motor drives a fan, the torque of which varies as the square of the speed. When running at 600 r/min it takes 28 A. Determine the resistance to be added in series with the armature to reduce the speed to 400 r/min. Neglect motor losses.

4.9 A 220-V dc shunt motor has an armature resistance of 0.3 Ω. Calculate (a) the resistance required in series with the armature to limit the armature current to 80 A at starting, and (b) the value of the counter EMF when the armature current has decreased to 30 A with the resistor still in the circuit.

4.10 A 15-hp 230-V shunt motor has a full load armature current of 56 A at 1200 r/min. The armature resistance is 0.2 Ω. Determine: (a) the internal developed power and torque at rated load; (b) the torque when the field current is reduced to 90% of its value instantaneously; and (c) the final speed for part (b). Assume that the horsepower load on the shaft remains constant throughout.

4.11 A 10-hp 200-V shunt motor has a full load efficiency of 85%. The armature resistance is 0.25 Ω. Determine the value of the starting resistor to limit I_{start} to $1.5 \times I_{full\,Load}$. Neglect the shunt current. What is the value of the counter EMF when the current drops to full-load value, assuming the starting resistor is still in the circuit?

4.12 A dc motor drives a 50-kW generator whose efficiency is 89.5%. (a) Determine the full-load kW rating of the motor. (b) If the overall efficiency of the motor-generator set is 78%, what is the efficiency of the motor? (c) What are the losses in each machine?

4.13 A 100-kW belt driven shunt generator runs at 600 r/min on a 250-V bus. When the belt breaks, it continues to run as a motor (supplied by the bus), then taking 8 kW. What will be its speed? Armature resistance is 0.02 Ω, and the field resistance is 125 Ω.

4.14 Figure 4.19 shows a dc motor under full-load condition driving a pulley mechanism to hoist an elevator at constant speed. (a) What is the minimum-size motor you would recommend? (b) If the line voltage V is reduced to 160 V, what would be the speed? (c) With $V = 120$ V, $R_F = 440$ Ω, and half-load, what would be the speed?

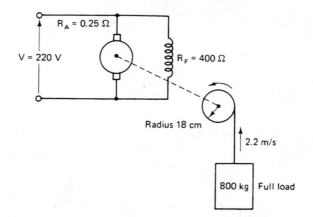

FIGURE 4.19.

4.15   A dc series motor drives an elevator load that requires a constant torque of 180 N·m. The dc supply voltage is 250 V, and the combined resistance of the armature and series field windings is 0.65 Ω. Neglect rotational losses and armature reaction effects. (a) The speed of the elevator is controlled by a solid-state chopper. At 60% duty cycle (i.e. $\delta = 0.6$) of the chopper, the motor current is 36 A. Determine the speed and the horsepower output of the motor and the efficiency of the system.

4.16   In Problem 4.15, if instead the elevator speed is controlled by inserting a resistance in series with the armature of the series motor, what resistance value would be needed to attain the speed obtained? Calculate the efficiency of this system, and compare with that obtained in Problem 4.15.

4.17   The speed of a 5-hp 250-V 900-r/min separately excited dc motor is controlled by a single-phase half-wave converter (see Figure 4.16). The armature resistance $R_A = 0.42\Omega$. The ac supply voltage is 220 V. The motor voltage constant is $k_g\phi = 0.22$ V/r/min. For a firing angle $\alpha = 30°$ and armature current of 20 A, determine the

    a.   motor speed;

    b.   motor torque; and

    c.   power to the motor.

Electric Power Generation

5.1 Introduction

The three-phase synchronous generator is used exclusively to generate bulk power. The voltage levels at which this power is generated is typically in the range 13.8 to 28 kV. As we will see, this voltage range is limited by practical insulating considerations. In addition, the number of conductors that can be placed into the stator slots presents further limitations. Conductors must have sufficient or adequate cross section to carry the current in order to keep ohmic losses to a practical minimum.

At the voltage levels mentioned it is possible to transmit the generated power directly, but this would result in unacceptable power losses and voltage drops, even if reasonably small distances are involved. It is therefore absolutely essential to distribute power at as high a voltage level as is practically feasible.

Transmission voltages vary routinely in the many hundreds of thousands of volts, up to the 760-kV level. This is made possible by power transformers. They transform the generated voltage at which transmission becomes feasible for distances well in excess of hundreds of miles. At the receiving end, this high voltage is transformed back to moderately high voltage levels, after which it is distributed to "local" consumers and users.

In studying alternating-current (ac) machines, starting with the synchronous generator, our objective is to analyze a machine in sufficient detail to develop an equivalent electric circuit which for practical purposes describes that machine. Such a circuit can then be used to predict that machine's performance without too many detailed calculations.

As before, the equivalent circuit will have an induced voltage depending on the air gap flux. It will also have reactances and resistances as part of the ac parameters at a particular frequency. We will deal with ac quantities now, not dc values. This inevitably leads to phasor diagrams representing the various voltage and current relationships as a visual aid in understanding machine performances. In this chapter we introduce some of these concepts, introduce the reader to the overall power system, and examine some synchronous generator constructional details.

Electric Power Systems

As was mentioned above, when power is transmitted at the generated voltage level, we do not "get too far." Let us investigate this further. Assume we have a 500-MW 22.5-kV three-phase synchronous generator. When delivering this rated power at an assumed unity power factor ($\cos\theta = 1.0$) results in a line current according to the expression for three-phase power ($P = \sqrt{3}\,V_L\,I_L\,\cos\theta$) of

$$I_L = \frac{P}{\sqrt{3}\,V_L\,\cos\theta} = \frac{500,000\text{ kW}}{\sqrt{3} \times 22,500 \times 1} = 12.83 \text{ kA}$$

These three currents will be carried in three identical bare overhead aluminum-cable steel-reinforced (ACSR) conductors, as shown in principle in Figure 5.1.

Assume that the conductors have a cross-sectional area A of 16 cm^2 (= 0.0016 m^2), which amounts approximately to a circular conductor having a diameter of 1.8 in. Let the

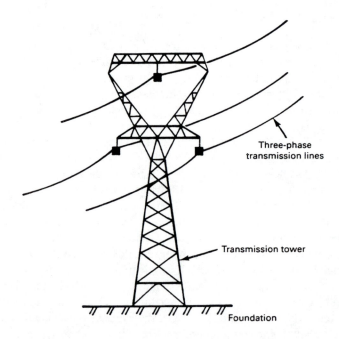

Three-phase
transmission lines

Transmission tower

Foundation

FIGURE 5.1. Overhead transmission line (500kV system).

conductor material be aluminum (resistivity $\rho = 0.280 \times 10^{-7}$ $\Omega \cdot$ m). From physics we have the equation, $R = \rho l / A$, where $l =$ conductor length in meters. This equation expresses the resistance of the conductor in terms of its physical dimensions l and A and its material property ρ. Let us further assume that we want to transmit this power over a distance of 20 miles (32.2 km). Substituting our values results in a conductor resistance (per line) of

$$R = 32,200 \times \frac{0.28 \times 10^{-7}}{0.0016} = 0.564 \ \Omega$$

We can now compute the ohmic loss per conductor as

$$P = I^2 R = (12,830)^2 \times 0.564 = 92.84 \ \text{MW}$$

which amounts to

$$\frac{92.84 \times 10^6}{32,200} = 2883 \ \text{W/m per conductor}$$

Thus 92.84 MW/conductor gives a total loss of $3 \times 92.84 = 278.5$ MW for our three-phase system. This amounts to a power loss of

$$\frac{278.5}{500} \times 100 = 55.7\%$$

of the generator output. In other words, more than one-half of the generated power is lost as heat in the conductors. Looking at the corresponding ohmic voltage drop in the line gives

$$V = IR = 12.83 \ \text{kA} \times 0.564 = 7.24 \ \text{kV/phase}$$

That is a rather large voltage drop. It becomes much worse if we consider line reactances. For a typical line its reactance is normally much larger than its resistance value. Typical reactance to resistance ratios for overhead lines vary from approximately 3 (at low voltages) to 20 or more (at ultrahigh voltages). This clearly illustrates that transmitting power under these conditions is absurd. The reader can easily verify that if the voltage was stepped up to say, 500 kV, the corresponding power losses drop to "negligible" proportions. The line current at this voltage level is "only" 577 A.

To raise the generator voltage to levels suitable for power transmission, we need a power transformer. Power transformers are always designed for a single frequency, for our purposes it is 60 Hz. Power transformers by definition are those with a rating of 250 kVA and up. For large power transformers we typically think in terms of thousands of kVA and more, up to 1,000 MVA. Of course, at the end of a transmission line we need another transformer to step down the voltage. Generators normally operate in parallel, and various generating stations supply the power grid. Referring to Figure 5.2, we see an overview of a typical electric power system. As can be seen, there are really four separate aspects to such a system:

1. Generation.
2. Transmission.

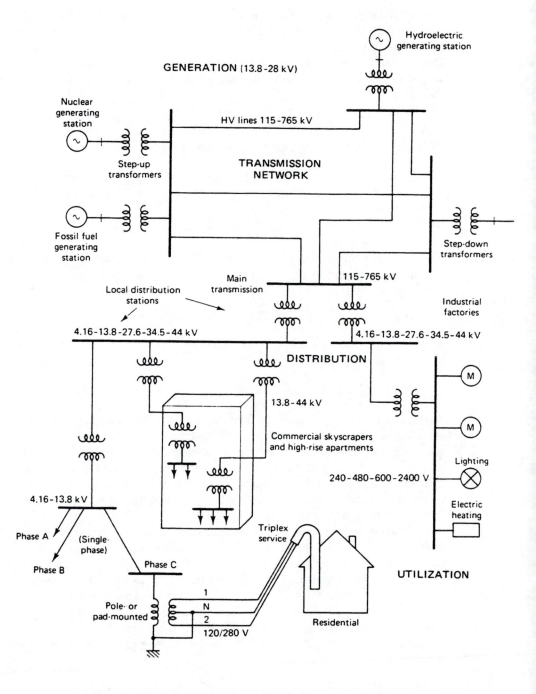

FIGURE 5.2. **Overview of a typical power system.**

3. Distribution.

4. Utilization.

In this book we confine ourselves to the generation of power and its utilization. That is, we want electric power to drive our machines to do a specific task. To this end we need to know what machines are available and what they can do, so that an educated choice can be made as to which machine we choose for a specific task. However, it is informative to see how the generated power at one end of the power system ends up at the other. Observe the various voltage levels and the role of the transformers. As mentioned, transformers with ratings over 250 kVA are *power transformers*. Those with ratings less than that are commonly called *distribution transformers*. They are usually located in the various power distribution stations at the end of the transmission system, which further distributes the power to localized centers.

5.2 Construction of Synchronous Generators

A synchronous generator could be built by replacing the cummutator in a dc machine by slip rings. The internal-generated ac voltage would then appear across the stationary brushes riding on these slip rings. For three-phase generated voltages we would need three slip rings, which must be connected to appropriate points on the armature winding.

In practice this construction is never used for commercial synchronous generators. Instead, a more dramatic change is employed, that makes the synchronous machine appear physically different from dc machines. Figure 5.3 shows the synchronous generator schematically. Unlike the dc generator, the synchronous generator must be driven at a constant speed. The reason is that the frequency of the generated voltage, hence that of the electrical network it supplies, is directly related to that speed. Thus the mechanical speed of the generator must be synchronized with the electrical frequency, from which the name *synchronous machine*.

The principle involved is simply this: Generator action depends entirely on the relative motion of the conductor with respect to the field lines. This suggests it is possible to construct an ac generator in which the role of the stationary and rotating member is interchanged, as compared to the dc machine. That is exactly what is done in practical generators, as Figure 5.3 illustrates. The winding in which the voltages are induced, the armature winding, is placed on the stator. The field circuit is placed on the rotor. This arrangement is preferable for several reasons:

1. The armature winding, being generally rated for high voltages and currents, is much larger and more complex than the field winding. Therefore, it can be better secured in the stationary member, both from a mechanical and electrical point of view.

2. The armature winding is easier to cool when stationary than when rotating. Because the stator core is larger, it makes it possible to provide better forced air circulation by being able to provide more air ducts. Forced cooling of the armature with circulating coolants, for example, water (necessary in large generators), is practical only in this type

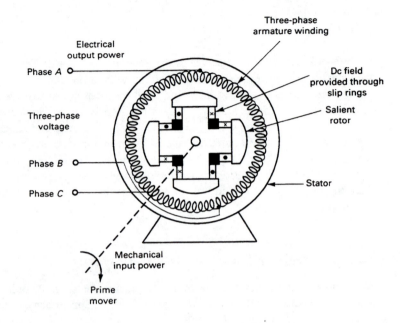

FIGURE 5.3.   Schematic representation of a four-pole synchronous generator showing reversal of the field and armature circuit as compared to a dc machine.

of construction. For extremely large generators 500 to 1000 MVA, enclosed hydrogen gas cooling is employed in addition to water-cooled stators. The water flows through the conductors and of course is separate from the hydrogen cooling system, which has a heat exchange system.

3. The field coils carry relatively small currents compared to the armature circuit. Their rotating electrical connections are thus smaller. No polarity reversals are necessary, and slip rings are usually employed. Also, slip rings through which the field receives current are inherently less prone to failure than are segmented commutators.

4. No commutator action is required, making the high power armature connections easier to make on a stationary member.

This explains why synchronous generators are different from dc machines. Not only that, synchronous generators are built in much larger sizes than dc generators. This is possible because the serious limiting factor of commutation on dc machines is not present on ac machines. Also, to meet the ever-growing demand of electric power, designers have kept pace by producing ever-larger sizes.

As an example to give some typical values, a certain 430,000-kVA 18.2-kV three-phase 1800-r/min 60-Hz synchronous generator has a rated armature current of 13,640 A. The field circuit will take 1780 A when working from a 500-V dc source. It would be impractical to conduct 13,640 A at 18,200 V (line-to-line) through slip rings and brush arrangement. The field circuit current, on the other hand, although substantial, has a working voltage of only 500 V and needs only two slip rings. Insulating difficulties between slip rings and shaft are thus greatly reduced. Furthermore, the high-voltage armature conductors are easier insulated on the stationary member. Last but not least, when generator sizes exceed 200 MVA, forced cooling of the armature conductors with liquid coolants is only practical with stationary windings.

These are the reasons for the inverse arrangement in construction as compared to dc machines. We will now take a closer look at some of the details of synchronous generator constructions.

Stators

As is typical in all electrical apparatus, the stator of generators consists of good electrical quality steel, laminated to minimize eddy current losses. By good electrical steel we mean that both the permeability and the resistivity of the material are high. Silicon steel meets this criterion. Figure 5.4 shows the stator slotted to receive the armature winding in much the same way as is done in dc machines. The number of slots is generally such that a symmetrical three-phase winding can be used. This is possible when the number of slots divided by the number of poles, times the number of phases, is an integer, that is,

$$\frac{\text{slots} \times \text{phases}}{\text{poles}} = \text{integer}$$

A brief discussion on the armature winding will be presented in Section 5.5. In slow-speed large-diameter machines such as hydroelectric generators having many poles, the axial length of the stator core is relatively short. In high-speed machines such as those driven by steam turbines, only two or four poles are generally used. The axial length of the core is many times the diameter in those machines. Why is this so? This ties in with the space the poles take up. The more poles that have to be fitted in, the larger the diameter. For example, a 60-Hz 225-r/min hydro unit will have 32 poles. But why the shorter length, then? This can be explained in terms of Eq. 2.1. This equation shows that the generated voltage is related to the radius R ($v = \omega R$) of the armature and the length L of the conductor in the field. The conductor length more or less determines the axial length of the machine. Therefore, we see that the parameters R and L determine the volume of the machine. For a specific machine power output, its size does not change much. It implies that for a certain rating (volume) making the diameter larger (to fit all the poles in) allows the length of the machine to be correspondingly shorter.

As the machine rating goes up, more armature copper is required. To accommodate the larger conductors, deeper slots are required. With deeper slots the outer diameter is approached and the teeth get wider (see Figure 5.5). This makes for mechanically stronger teeth on the stator. In addition to the advantages already listed for inverting the power circuit, we have an added bonus here. Because if slots on a rotating armature are made deeper, we approach the center of that armature and the teeth become narrower and mechanically weaker, as shown in Figure 5.5.

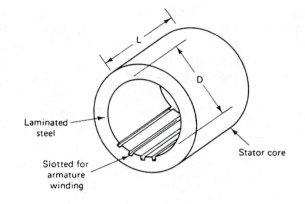

FIGURE 5.4.   Slotted and laminated stator core.

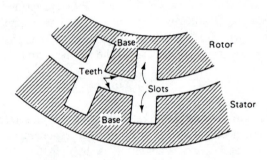

FIGURE 5.5.   For equal slot depth teeth on stator become wider and stronger at the base.

About 55% of stator circumference must be provided for teeth to carry the necessary magnetic flux safely without exceeding maximum flux density. This leaves about 45% of the circumference for slots to be filled with conductors and insulation. More coils means a greater number of slots. On the other hand, fewer but wider slots means fewer turns of heavier conductors. In the first case a higher voltage with lower current rating may be in order. In the second, we may aim for a higher current rating but lower voltage.

This leads next to ratings. The generator output is limited by its magnetic capability. Like all magnetic devices, they are subject to iron saturation. As magnetic flux densities increase, a point is reached where any increase in excitation current results in little change in magnetic flux. This normally happens at the point where the generator voltage is at its maximum. Practical operation takes place somewhere below this point.

Machine ratings are normally given in kVA (kilovolt-amperes). The volt-ampere rating, generally lower than magnetic capability, is that load where the design temperature has been reached. Exceeding this rating means the machine gets too hot and the insulation deteriorates faster. Rating determines overall size, which as we stated above, relates to machine dimensions, a given design type, cooling method, and so on. In practice, every machine has its rating specified on a nameplate attached to it, specifying the output power it is designed to provide. In addition, the manufacturer's name appears on it, as well as current and voltage ratings. At nominal operating conditions, the output power of the machine is referred to as full-load power, rated power, or nameplate power.

Rotors

The rotating member of a synchronous generator is normally constructed in one of two ways: (1) with salient or projected poles or, (2) with a round rotor or cylindrical rotor. Figure 5.6 indicates these rotor types.

The salient rotor machine has dc current in the rotor field coils that provide the MMF $(NI)_{field}$ to set up the magnetic field shown. Because the pole faces are tapered, flux path l_2 contains a wider air gap than l_1. As discussed in Chapter 1, a larger air gap means more reluctance. As a result, the flux density of path l_2 is less than that of l_1.

As mentioned, the pole shoe is shaped to make the resultant flux density waveform in the air gap sinusoidal. The result is then a generated voltage that is also shaped sinusoidally.

Considering the cylindrical rotor of Figure 5.6b, the dc winding is placed in rotor slots. The air gap is uniform, but we see that path l_1 encloses more conductors carrying current than path l_2. Therefore, there exists a greater MMF along path l_1. Again a better sinusoidal flux density distribution is obtained.

The salient pole construction is generally restricted to low-speed generators, such as those driven by water wheels. Because of their low speeds they require a large number of poles, as Eq. 5.1 will verify, assuming a fixed frequency. For example, we need 48 poles for a 150-r/min 60-Hz generator. The salient pole structure is simpler and more economical to manufacture than a corresponding cylindrical rotor type. Besides, if the number of poles becomes large, the cylindrical rotor construction becomes impossible to construct in practical machines. Therefore, cylindrical rotors are exclusively used for generators driven by steam turbines, as such, they are known as *turboalternators* or *turbogenerators*. They are usually two or four poles because they are used for high-speed applications. Since these rotors are compact, they readily withstand the centrifugal forces developed in large size generators at those speeds. As Figure 5.6b shows, the field winding is placed in slots that run parallel to the rotor axis. It is the location of the slots creating the magnetic poles, since physically there are no poles to be seen. Like the stator, the salient pole rotor is normally laminated. Cylindrical rotors are sometimes not laminated and are then made of a single steel forging. The slots are machined out, and the primary function of the rotor teeth is to hold the field coils in against the centrifugal forces. Heavy nonmetallic wedges are forced into the grooves in the teeth to close the slot.

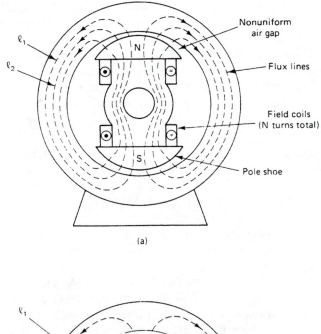

(a)

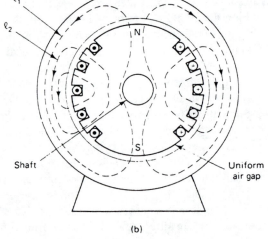

(b)

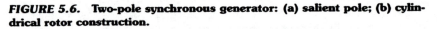

FIGURE 5.6. **Two-pole synchronous generator: (a) salient pole; (b) cylindrical rotor construction.**

5.3 *Three-Phase Voltage Generators*

Rotor speed and frequency of the generated voltage are directly related. Let us now consider how. Consider the elementary two-pole ac generator of Figure 5.7. To simplify things there is only one coil shown, made up of two conductors in series, conductor a and a'. When we have a single coil like this we talk of a *concentrated winding*. In real generators there are many coils in each of the three phases which are distributed in slots over the entire stator periphery. As such, the winding is known as a *distributed winding*. Naturally, each coil could have many turns, although only one turn per coil is shown here. Direct current is supplied to the field winding on the rotor through brushes and slip rings.

Assume the rotor is turned clockwise. This means that the rotor field flux moves with respect to the armature coil, a necessary condition for generating voltage. In fact, the condition is the same as Figure 2.4 predicts, except here the field moves. We can therefore expect the generated voltage as a function of angular rotor position (or time) also to be similar to that shown in Figure 2.5. Thus, turning the rotor one complete turn starting from the position shown generates one cycle of the voltage waveform. Rotating the rotor one turn in 1 s gives one cycle per second, or 1 hertz (Hz), as the frequency is referred to. Turning the rotor around two times in 1 s yields two cycles, and the frequency is 2 Hz. The number of cycles per second, or frequency, is thus directly proportional to the speed of the rotor. If the speed is 60 revolutions per minute (r/min), the frequency f equals 1 Hz, or $f = 1$ Hz. For a frequency of $f = 60$ Hz, the rotor must turn at 3600 r/min. For a rotor speed of n r/min, the rotor turns at a speed of $n/60$ revolutions per second (r/s). Because of the continuous nature of the magnetic field lines, magnetic poles always occur in pairs. If the rotor has more than two poles, say P poles, each revolution of the rotor induces $P/2$ cycles

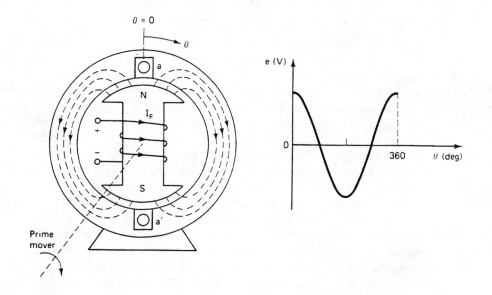

FIGURE 5.7. Schematic diagram of a single-phase two-pole ac generator and the generated voltage in coil $a - a'$.

of voltage in the stator coil. The frequency of the induced voltage as a function of rotor speed is then

$$f = \frac{P}{2}\frac{n}{60} = \frac{Pn}{120} \text{ Hz} \tag{5.1}$$

where f = generated frequency, Hz

n = rotor speed, r/min

P = number of poles on the rotor

For the generator to produce a sinusoidal voltage of a given frequency, the rotor must be turning at a speed compatible with the frequency. The rotor is said to be turning at synchronous speed if its speed corresponds to n r/min in Eq. 5.1.

EXAMPLE 5.1

A four-pole ac generator operates at 1800 r/min.
 a. What frequency does it generate?
 b. What must be the speed if the frequency is 50Hz?

Solution
 a. $f = Pn/120 = 4 \times 1800/120 = 60$ Hz.
 b. $n = 120f/P = 120 \times 50/4 = 1500$ r/min.

For a three-phase synchronous generator we must have three such coils, as shown in Figure 5.7. The two additional coils must be placed on the generator stator in such a way that the three coils are all 120 electrical degrees apart. The machine is then called a *three-phase synchronous* generator. Figure 5.8 shows the machine of Figure 5.7 with two additional coils added. Each coil is displaced by 120° degrees to each of the other two coils. Figure 5.9 shows the voltages produced. Curves e_A, e_B, and e_C show the instantaneous values of the phase voltages, while the effective or rms values E_A, E_B, and E_C are used in drawing the phasor diagram. The subscripts A, B, and C refer to the phase sequence of the voltages (i.e., the order in which they are generated). The principles described for the simple ac generator in Figure 5.7 also apply to the three separate single-phase voltages induced in the three separate coils of the generator in Figure 5.8. These three coils on the stator are usually connected either wye (Y) or delta (Δ), to produce a three-phase voltage source, as illustrated in Figure 5.10. The induced voltage in each stator coil is known as a phase voltage E, and the voltage appearing between any of two line conductors is known as the line voltage V_L, or the terminal voltage V_t when measured at the machine terminals.

Each of the phase voltages $\mathbf{E}_A$, $\mathbf{E}_B$, and $\mathbf{E}_C$ is completely specified by its magnitude, frequency, and phase angle. The magnitude of each phase voltage is

$$E_{max} = B_m\, l\, \omega\, r \quad \text{V} \tag{5.2}$$

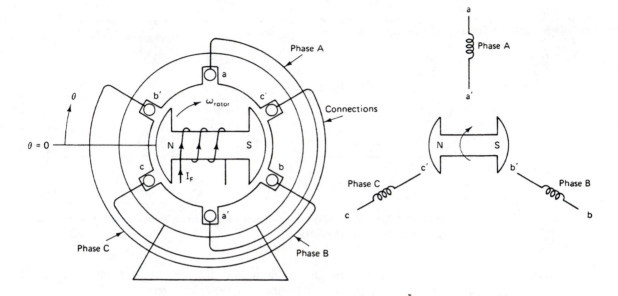

FIGURE 5.8.   Simple two-pole three-phase ac generator. Armature winding concentrated in one slot per phase per pole.

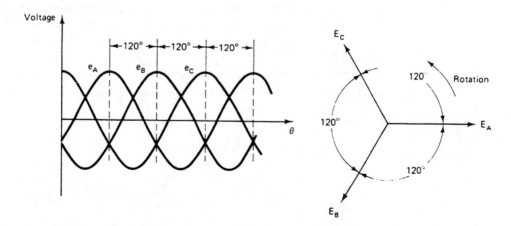

FIGURE 5.9.   Voltage waveforms and phasor diagram of the three-phase generator in Figure 5.8.

where B_m = maximum flux density produced by the rotor field winding, T
             l = length of both coil sides in the magnetic field, m
           ω = angular velocity of the rotor, rad/s (= $2\pi \times$ frequency, rad/s)
            r = radius of the armature, m

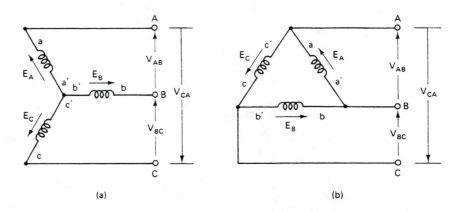

FIGURE 5.10. **Coil connections to form a three-phase voltage source: (a) Y-connected; (b) Δ-connected.**

Equation 5.2 is similar to Eq. 2.1. If the rotor is driven by its prime mover at a constant speed, the voltage can be adjusted by changing the field current.

EXAMPLE 5.2

The two-pole generator shown in Figure 5.8 generates a three-phase voltage of phase sequence ABC as illustrated in Figure 5.9. Assuming the following data: B_m =1.2 T, armature length is 0.5 m, the rotor is turned at 1500 r/min, and the inside diameter of the stator core is 0.4 m.

a. Determine the magnitude of the induced voltage per phase.

b. The expression of generated voltages in the time domain.

c. Express the voltages as phasors.

Solution

a. Equation 5.2 gives the magnitude of the generated voltage. Thus

$$E_{max} = B_m l \omega r$$
$$= 1.2 \times (0.5 + 0.5) \times \frac{2\pi \times 1500}{60} \times \frac{0.4}{2} = 37.7 \text{ V}$$

b. The generated voltages are sinusoidal. Taking phase A as the reference, we see phase B lags phase A by 120 electrical degrees (or $2\pi/3$ radians), and phase C lags phase A by 240 electrical degrees (or $4\pi/3$ radians). From Eq. 5.1,

$$f = \frac{2 \times 1500}{120} = 25 \text{ Hz}$$

which is the frequency of the generated voltage; therefore,

$$\omega = 2\pi f = \pi 50 = 157 \text{ rad/s}$$

Taking phase A as the reference voltage gives

$$e_A = E_{max} \sin \omega t = 37.7 \sin 157t \text{ V}$$
$$e_B = E_{max} \sin(\omega t - 2\pi/3) = 37.7 \sin(157t - 2\pi/3) \text{ V}$$

and

$$e_C = E_{max} \sin(\omega t - 4\pi/3) = 37.7 \sin(157t - 4\pi/3) \text{ V}$$

c. Representing the voltages as phasors, we need their rms values:

$$E_{rms} = \frac{E_{max}}{\sqrt{2}} = 26.7 \text{ V}$$

Then

$$E_A = 26.7 \angle 0° \text{ V}, \quad E_B = 26.7 \angle -120° \text{ V}, \quad E_C = 26.7 \angle -240° \text{ V}$$

EXAMPLE 5.3

For the generator in Example 5-2, calculate the line voltages if the armature winding is (a) Y-connected, (b) Δ-connected.

Solution

a. Y-connected.   From circuit theory or by application of Kirchhoff's law applied to Figure 5.10, we have

$$V_{AB} = E_A - E_B \quad V_{BC} = E_B - E_C \quad V_{CA} = E_C - E_A$$

We can represent this graphically (see Figure 5.11). Note that the phasors V_{AB}, V_{BC}, and V_{CA} form again a set of three-phase voltages. Their magnitudes are readily shown to be larger than their phase voltages by a factor of $\sqrt{3}$. Thus

$$V_{AB} = 26.7\sqrt{3}\angle 30° = 46.2\angle 30° \text{ V}$$

$$V_{BC} = 46.2\angle -90° \text{ V} \quad V_{CA} = 46.2\angle 150° \text{ V}$$

The phase relationship between the corresponding line voltages is identical to that of the phase voltages. Note, however, that the set of line voltages is shifted in phase by +30 electrical degrees from the set of phase voltages.

b. Δ-connected. By inspection of Figure 5.10b we see that the line voltages are identical to the phase voltages. Thus

$$V_{AB} = 26.7\angle 0° \text{ V} \quad V_{BC} = 26.7\angle -120° \text{ V} \quad V_{CA} = 26.7\angle -240° \text{ V}$$

which form a three-phase voltage set.

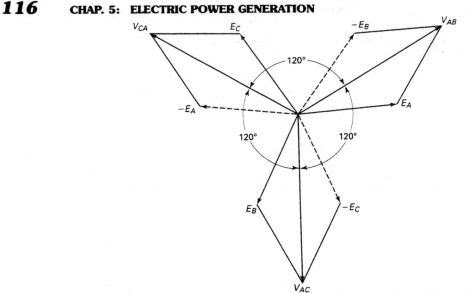

FIGURE 5.11.   Phasor diagram refers to Y connection in Example 5.3.

As will be evident as we proceed, for a specific machine rating connecting the armature in Y results in larger terminal voltages. The line currents, however, equal the phase currents. For a Δ-connected armature winding, the terminal voltages equal the phase voltages, but the line currents will be larger by the factor $\sqrt{3}$. In total, the power delivered by the generator is the same in both instances.

5.4   *Armature Windings*

There are many possible ways of winding an armature of synchronous generators. The majority of them come in two general types: (1) single-layer windings, and (2) double-layer windings. The material presented here deals only with the principles involved rather than all of the practices of armature winding construction.

As Figure 5.8 showed, a three-phase winding results by adding two more sets of armature coils to that of the generator shown in Figure 5.7. They are displaced 120 and 240 electrical degrees from the first coil (phase) to produce a system of three voltages equal in magnitude and displaced from each other by 120° (see Figure 5.9). The machine is than called a *three-phase generator*. Figure 5.12 illustrates a three-phase four-pole synchronous generator with a developed view of the armature winding. There is only one coil side in each slot, making it a single layer winding. If the windings of the three phases start at s_a, s_b, and s_c and finish at f_a, f_b, and f_c, they may be joined in two ways. These of course are the delta (Δ) and wye (Y, also called the *star connection*) connections, as illustrated in Figure 5.10. For a given number of turns per phase, the Y connection gives a higher terminal voltage than the Δ connection, but a correspondingly smaller output current.

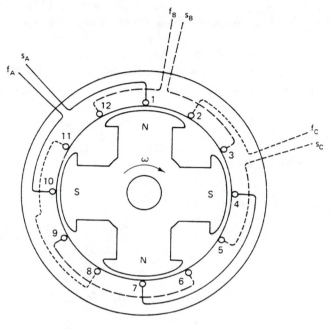

(a)

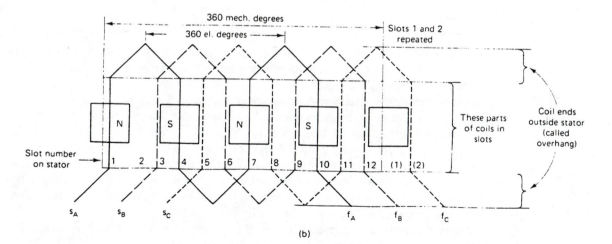

(b)

FIGURE 5.12. **Three-phase synchronous generator:** **(a) three-phase single-layer armature winding distributed in one slot per pole per phase; (b) developed view of the armature winding.**

Some observations can be made regarding the three-phase armature winding in Figure 5.12. The separation between the phase windings is 120 electrical degrees, or 60 mechanical degrees. This we conclude from simple reasoning that a full EMF cycle will be generated when the four-pole rotor turns 180 mechanical degrees. A full EMF cycle represents 360 electrical degrees. Extending this to a P-pole generator (P must always be a positive even integer), we have the following important relationship between mechanical rotor angles α_{mech} and electrical angles α_{el}:

$$\alpha_{el} = \frac{P}{2} \alpha_{mech} \tag{5.3}$$

EXAMPLE 5.4

A three-phase synchronous generator has 12 poles. What is the mechanical angle corresponding to 180 electrical degrees?

Solution   The mechanical angle between a north and south pole is

$$\alpha_{mech} = \frac{360 \text{ mechanical degrees}}{12 \text{ poles}} = 30°$$

This corresponds to 180 electrical degrees. Using Eq. 5.3 as a check results in

$$\alpha_{el} = \frac{P}{2}\alpha_{mech} = \frac{12}{2} \times 30° = 180°$$

as before.

The generated EMFs in each coil side aid each other in each phase to add up to the total phase voltage. This is readily verified by using the right-hand rule to the coil sides making up the phase windings in Figure 5.12.

Connecting the phases in Y as indicated in Figure 5.13, has the added advantage that the neutral point of the generator can be brought out. This means that not only line voltages are available, but the line to neutral (the phase) voltage as well. Furthermore, the generator neutral is normally grounded. It enables grounding of apparatus and devices at the consumer end of electrical energy, a very desirable feature from a safety point of view.

For the indicated direction of rotation of the rotor in Figure 5.12a (clockwise), the resulting phase sequence of the three-phase supply is ABC. This means that the maximum voltage is generated in Phase A, followed by Phase B and then Phase C. Reversing the direction of rotation would result in an ACB sequence, or *negative phase sequence*. The former (ABC) is often referred to as the *positive phase sequence*.

In summary, we have obtained a symmetrical set of three-phase generated EMFs:

$$\begin{aligned}
\mathbf{E}_A &= E_A \angle 0° \text{ V} \\
\mathbf{E}_B &= E_B \angle -120° \text{ V} \\
\mathbf{E}_C &= E_C \angle -240° \text{ V}
\end{aligned} \tag{5.4}$$

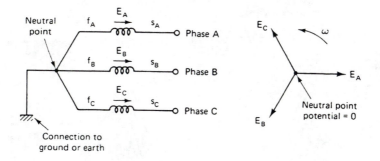

FIGURE 5.13. **Y-connected armature winding of Figure 5.12; phase sequence ABC.**

in which the A phase is taken as the reference phasor. The boldface type for the EMFs $\mathbf{E}_A$, $\mathbf{E}_B$, and $\mathbf{E}_C$ indicates phasor quantities; E_A, E_B and E_C in italic type are the magnitudes of those phasors making an angle with the reference axis as indicated.

When the three-phase windings $s_a - f_a$, $s_b - f_b$, and $s_c - f_c$ are interconnected as shown in Figure 5.13, it is readily verified that for a symmetrical three-phase EMF set, the following relationship holds:

$$\mathbf{E}_A + \mathbf{E}_B + \mathbf{E}_C = 0 \tag{5.5}$$

Double-Layer Windings

The armature winding of Figure 5.12 has only one coil side per pole per phase. In effect, each phase winding has only two coils in series. If the slots were not excessively wide, each conductor in a given slot would generate the same voltage. Each phase voltage would then be equal to the product of the voltage per conductor and the total total number of conductors per phase. In reality this would not be a very effective way of using the stator core, because of the variation in flux density in the iron and the localized heating effects in the slot regions. This would lead to a flat-topped waveform instead of sinusoidal, resulting in harmonics. To overcome this problem, practical generators have the winding distributed in several slots per pole per phase. Figure 5.14 shows a section of an armature winding and the general type of coil used in practice. The winding is distributed in two slots per pole per phase. Note that there are 24 slots now in the stator core as opposed to 12 slots shown in Figure 5.12. There are two coil sides per slot, and each coil has more than one turn. All coils are of the same shape, which means they can be preformed prior to assembly. The shape of a typical coil is shown in Figure 5.15.

As seen in Figure 5.14, one side of a coil is placed on top of another belonging to the same phase. All coils so placed on the stator fit snugly together, as the photographs in Figures 5.16 and 5.17 attest. Those parts of the coils not located in slots are generally referred to as the *winding overhang*. Since no voltage is introduced in the overhang (as it is not in the magnetic field), it is kept as short as possible. Naturally we cannot avoid the overhang altogether since the coil sides must be connected together to form a coil.

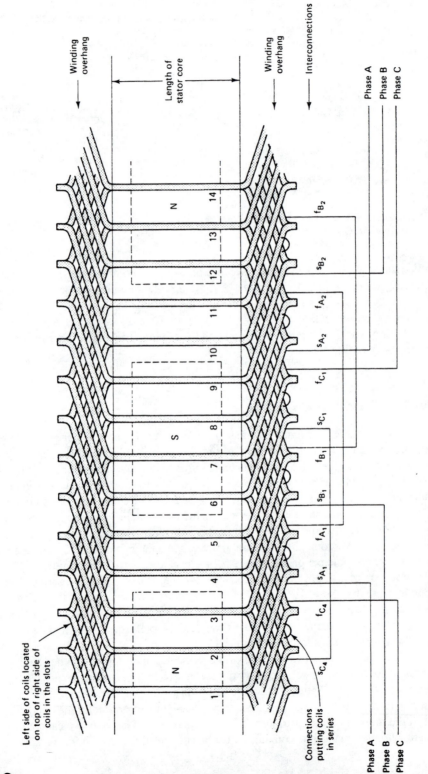

FIGURE 5.14. Section of a double-layer armature winding in a four-pole, three-phase synchronous generator. Winding is distributed in two slots per pole per phase.

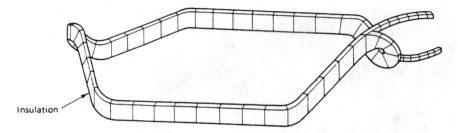

FIGURE 5.15.   Multiturn coil of a double-layer armature winding.

Insulation

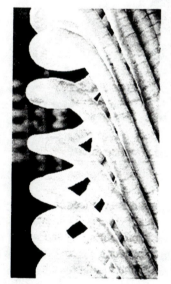

FIGURE 5.16.   Section of the overhang of a stator winding. (Courtesy of Siemens.)

Distribution Factor

When a winding such as the one we are discussing is made up of a number of coils placed in separate slots, the EMFs generated in the various coils per phase are not in phase. Consequently, the terminal EMF is less than if the winding had been concentrated. The factor by which the EMF of a distributed winding must be multiplied to give the total generated EMF is called the *distribution factor* k_d for the winding. This factor is always less than unity.

Assume there are n slots per phase per pole. The spacing between slots in electrical degrees is then

FIGURE 5.17. View of a synchronous generator stator. (Courtesy of Siemens.)

$$\Psi = \frac{180 \text{ electrical degrees}}{n \times m} \tag{5.6}$$

where m is number of phases.

Referring to Figure 5.18a, it is apparent that the induced EMF in slot 2 will lag behind the induced EMF in slot 1 by $\Psi = 15$ electrical degrees. The EMF induced in slot 3 will be 2 Ψ degrees behind in phase, and so on. Representing these EMFs by phasors $\mathbf{E}_1$, $\mathbf{E}_2$, $\mathbf{E}_3$, and $\mathbf{E}_4$, respectively, one would obtain the phasor diagram shown in Figure 5.18b. The total stator EMF per phase $\mathbf{E}$ is then the vector sum of all phasors, namely,

$$\mathbf{E} = \mathbf{E}_1 + \mathbf{E}_2 + \mathbf{E}_3 + \mathbf{E}_4$$

It is clear because of the displacement angle Ψ, the total stator EMF E is less than the arithmetic sum of the coil EMFs by the factor

$$k_d = \frac{\text{Vector sum}}{\text{arithmetic sum}} = \frac{\mathbf{E}_1 + \mathbf{E}_2 + \mathbf{E}_3 + \mathbf{E}_4}{4 \times |\mathbf{E}_{coil}|}$$

and k_d is called the distribution factor. It can be shown that

$$k_d = \frac{\sin(\frac{1}{2}nm)}{n \, \sin(\Psi/2)} \tag{5.7}$$

with the parameters as defined before.

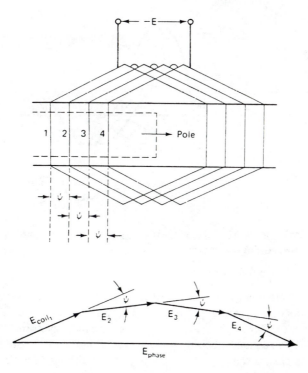

FIGURE 5.18. Phasor diagram of induced voltage in a group of four coils in a distributed winding.

The real advantage of a distributed winding, however, is the improvement in generated voltage waveform. Consider Figure 5.19; for simplicity three coils per group are chosen (i.e., there are three coils in series per phase for every pair of poles). This means there are 20 electrical degrees between successive coils.

To dramatize the point we are trying to make, it is assumed that each coil has a generated waveform which is nonsinusoidal, as illustrated in Figure 5.19. Adding the three-component EMFs point by point gives a resultant waveform e_t. Observe the resultant waveform, which approximates a sinusoidal waveform fairly closely. The waveform will improve even more as the number of component waves is increased. Further note that the maximum value of the resultant waveform is somewhat less than three times the magnitude of the coil EMFs, due to the distribution factor.

Fractional Pitch Coils

In a full-pitch coil the generated voltage in both coil sides are in phase. In practice the distance between the two coil sides is usually made less than full pitch. The whole idea

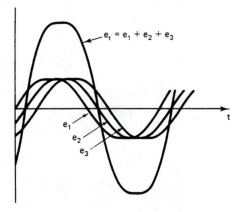

FIGURE 5.19.   Total EMF e_t, of three nonsiusoidal EMFs, becomes nearly sinusoidal upon addition.

is to ensure that the generated EMF is sinusoidal. This is easier accomplished by using fractional-pitch windings. In fractional-pitch coils, the coil sides are out of phase, the number of EMFs to be added is doubled, and the waveform is improved.

From a practical point of view, in machines with six or fewer poles it is often impractical to place a full-pitch coil into a machine without damaging the coil because excessive deformation of the coil form is required to fit the winding in. Fractional-pitch coils normally have shorter end connections than full pitch coils; therefore, there is some saving in copper and weight.

There is a disadvantage of using fractional-pitch coils, and that is a slight reduction in generated EMF as compared to full pitch coils. It is customary to express this reduction in voltage by a *pitch factor*, k_p. Its value may be calculated from the equation

$$k_p = \sin \frac{p^o}{2} \tag{5.8}$$

where p^o is span of the coil in electrical degrees.

EXAMPLE 5.5

A synchronous generator has 12 poles, and a three-phase stator winding placed in 144 slots; the coil span is 10 slots. Determine the distribution factor k_d, the pitch factor k_p, and the winding factor $k_w = k_d \times k_p$.

Solution   The number of slots per pole per phase

$$n = \frac{\text{slots}}{\text{phases} \times \text{poles}} = \frac{144}{3 \times 12} = 4$$

The electrical angle between slots

$$\Psi = \frac{180}{n \times m} = \frac{180}{4 \times 3} = 15°$$

Then from Eq. 5.7

$$k_d = \frac{\sin(\frac{1}{2}n\Psi)}{n \sin(\Psi/2)} = \frac{\sin(\frac{1}{2} \times 4 \times 15)}{4 \times \sin(15/2)} = 0.958$$

The pole pitch is 144/12 = 12 slots. The coil span in electrical degrees is

$$p = \frac{\text{coil pitch}}{\text{pole pitch}} \times 180° = 150 \text{ electrical degrees}$$

Then, according to Eq. 5.8,

$$k_p = \sin\frac{150°}{2} = 0.966$$

The *winding* factor

$$k_w = k_d \times k_p = 0.958 \times 0.966 = 0.925$$

which is the factor by which the generated voltage is reduced.

The minor disadvantage of the distributed winding with short pitch coils is a somewhat reduced voltage as expressed by the winding factor. However, the many advantages gained, such as a uniform coil shape on the winding and a purely generated sinusoidal voltage, far outweigh this disadvantage.

5.5 Excitation Systems

AC generators have to be excited with dc current to provide the rotor field—they cannot be self-exciting. The excitation current is often supplied by a small dc generator, called an *exciter*. The exciter can be mounted on the end of the synchronous generator shaft or can be driven by a separate motor. The excitation is fed to the rotor field winding via brushes and slip rings. Another arrangement is to use a solid-state power supply (rectifier) energized by the generator terminal voltage. Still another method is the *brushless exciter*, where the excitation current is obtained from a separate ac winding placed on a separate rotor, but connected to the main rotor. The ac voltage is then rectified by solid-state rectifiers placed on the rotor. This system is indicated in Figure 5.20. The system is "self-exciting" and needs no outside power source. Brushless excitation systems can be built up to 700 kW to control a 1000 MVA turbine generator. Absence of brushes and commutator should be noted, making the unit more reliable. Typical dc voltage levels for the field circuit are multiples of 125 V, up to 500 V.

FIGURE 5.20. **Stator and cylindrical rotor of a synchronous generator, with rotor of brushless exciter with rotating rectifiers. (Courtesy of Siemens.)**

PROBLEMS

5.1   At what speed must an eight-pole synchronous generator rotate in order to generate a voltage with a frequency of 60 Hz?

5.2   A dc motor is used as a prime mover for a synchronous generator in order to obtain a variable-frequency supply. If the speed range of the motor is 820 to 1960 r/min and the generator has four poles, what is the obtainable frequency range of the output voltage?

5.3   A 1000-kVA 2200-V 60-Hz three-phase synchronous generator is Y connected. Calculate the full-load line current.

5.4   If the generator in Problem 5.3 delivers a load of 720 kW at a power factor of 0.80, calculate the line current.

5.5   A three-phase load of 10 Ω/phase may be connected by means of switches in Y or Δ. If connected to a 220-V three-phase synchronous generator, calculate

    a.   the power dissipated in Y connection, and
    b.   the power dissipated in Δ connection.

5.6   A 250-kVA 1260-V Y-connected synchronous generator has its armature winding reconnected in Δ. Determine the machine rating when Δ-connected (i.e., kVA, I_L, and V_L).

5.7 The phase voltage of a 60-Hz synchronous generator is 4600 V at a field current of 8 A. Determine the open-circuit voltage at 50 Hz if the field current is 6 A. Neglect saturation.

5.8 If the generator of Problem 5.7 is used to generate 50-Hz voltage, what will be the line voltage if the armature is Y-connected? Δ-connected?

5.9 A four-pole synchronous generator is driven at 1800 r/min and generates a three-phase voltage of sequence ACB. Assume the following machine data: $B_m = 1.2$ T, armature length 0.5 m, inside stator core diameter 0.4 m.

    a. Determine the magnitude of the induced voltage/phase.
    b. Express the generated voltages in the time domain.
    c. Express the voltages as phasors.

5.10 For the generator in Problem 5.9, calculate the line voltages if the armature winding is Y-connected.

5.11 Using the three-phase voltage set in Eq. 5.4, prove Eq. 5.5.

5.12 Determine the distribution factor of a three-phase eight-pole synchronous generator winding having 120 slots on the stator.

5.13 If the armature winding in Problem 5.12 has a coil span of 12 slots, calculate the pitch factor. What would be the winding factor for this armature winding?

5.14 A 60-Hz three-phase hydroelectric generator has a rated speed of 120 r/min. There are 558 stator slots with two conductors per slot. The air gap dimensions are $D = 6.1$ m and $L = 1.14$ m. The maximum flux density $B_m = 1.2$ T. Calculate the generated voltage per phase.

5.15 A six-pole synchronous generator has a total of 108 slots. Determine

    a. the number of slots per pole,
    b. the electrical and mechanical angle between adjacent slots, and
    c. the distribution factor.

6

Synchronous Generators

6.1 Introduction

In ac machines, to which the remainder of the book is devoted, the fluxes are not constant; in fact they vary or move about continually. We have already seen in Chapter 5 that the field winding of a synchronous generator is placed on the rotating member, the rotor. As with most ac machinery, the armature winding is placed on the stationary member, the stator, so that the power circuit is stationary. Large-scale power generation, that is, the conversion of mechanical energy to electrical form, is done by synchronous generators. Electric utilities generate three-phase power exclusively. Therefore, only three-phase synchronous generators are discussed in this chapter. Construction details of the generator were discussed in Chapter 5; now we concentrate on its performance characteristics. Some three-phase concepts necessary to an understanding of three-phase machines were explained earlier; others will be presented as we go along.

6.2 Magnitude of the Generated Voltage

We already encountered an expression for the voltage as given by Eq. 5.2. That expression is not used in practice because the value of r is not readily known. Therefore, an alternative relationship is desired. As expected, the average induced voltage in each phase winding as the rotor field sweeps by is governed by *Faraday's law*. It is in formula form as follows:

$$E_{avg} = N\frac{\Delta\phi}{\Delta t} \text{ V}$$ (6.1)

where E_{avg} = average generated voltage in the winding, V
$\Delta\phi$ = change of flux in a given time, Wb
Δt = the time in which this flux change takes place, s
N = number of turns on the phase winding

Figure 6.1 illustrates such a coil of the armature winding while a field pole moves past. It should be noted that it is the relative motion between the field and the coil that counts, so that if the coil is moving from left to right with the poles stationary, identical results are obtained.

In Figure 6.1, the flux change between the two positions shown is ϕ_m, in webers. This implies a pole movement equal to one-half of the pole pitch in which one-fourth of a voltage cycle is generated. Since one cycle occurs in $1/f$ seconds, the elapsed time for this part of the waveform will be $1/(4f)$ seconds. Therefore, substituting this value of elapsed time in Eq. 6.1 gives

$$E_{avg} = N\frac{\phi_m}{1/(4f)} = 4fN\phi_m \text{ V}$$ (6.2)

Equation 6.2 is a general expression applicable to all generators regardless of the pole-flux distribution. Naturally, if the flux distribution is sinusoidal the generated voltage will be sinusoidal. If so, we know from circuit theory that the effective or rms value of the voltage is 1.11 times the average value. Thus

$$E = 4.44fN\phi_m \text{ V}$$ (6.3)

where E = generated rms voltage in each phase winding, V
f = frequency of the generated voltage, Hz
ϕ_m = maximum flux per pole, Wb

Practically, the flux distribution may not be a perfect sine wave and the effective voltage is slightly modified to account for this. Recall that the factor 1.11 (the *form factor*, as it

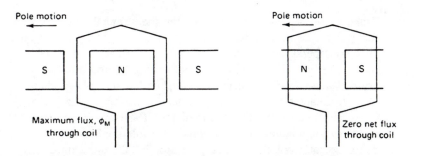

FIGURE 6.1. Illustration of flux change through a coil as field poles move past the coil from right to left.

is called) is the ratio between effective and average voltage and applies only to sinusoidal waveforms.

EXAMPLE 6.1

Determine the effective voltage generated in one of the phases of a synchronous generator. The following data is given: $f = 60$ Hz, turns per phase $N = 230$, and maximum flux per pole $\phi_m = 0.04$ Wb.

Solution

$$
\begin{aligned}
E &= 4.44\,fN\phi_m \\
&= 4.44 \times 60 \times 230 \times 0.04 = 2450 \text{ V/phase}
\end{aligned}
$$

For a Y-connected three-phase generator, this results in an open-circuit line-to-line (or terminal) voltage of

$$
V_L = 2450 \times \sqrt{3} = 4244 \text{ V}
$$

6.3 Voltage Regulation

The principles of the three-phase synchronous generator have been discussed and we have seen how a three-phase voltage source is obtained. Before the generator can supply an electrical load, it must fulfill the following requirements:

1. It must operate at the correct or synchronous speed as it determines the frequency, since the number of poles is fixed for any given machine.
2. The dc rotor field excitation must be present.
3. It must have the correct output voltage, which can be set by adjusting the field excitation current.

When a load is placed on the generator, the terminal voltage will be affected despite the fact that the dc field excitation remains constant. The way in which the terminal voltage changes depends on the character of the load, that is, on its power factor. Resistive and inductive loads will cause the terminal voltage to drop by as much as 25 to 50% below the no-load value, whereas a capacitive load will tend to raise it above the no-load value.

In Chapter 3, where the dc generator was discussed, it was shown that two factors are responsible for the change in voltage. These are the armature resistance voltage drop and the change in flux due to armature reaction (see Section 2.5). As we may expect, these factors are also present in an ac generator, but there is a third effect as well: the armature reactance voltage drop. This voltage drop is caused by the inductance L_A of the armature winding, which is considerable. It asserts itself as a reactance X_A, where

$$X_A = 2\pi f L_A \quad \Omega \tag{6.4}$$

in which L_A is the armature winding inductance per phase, in henries, and $\omega = 2\pi f$ is the radial frequency, in radians per second.

The reactance of the armature then gives rise to an additional voltage drop. This must be accounted for to arrive at the generated voltage. As will be shown later, the variation in flux due to the armature reaction effect is generally also treated as a voltage drop. This will be discussed in more detail in the following sections. Before proceeding let us refer to Figure 6.2, which depicts typical generator terminal characteristics. The terminal voltage is presented as a function of the load current at various load power factors; the field excitation current is constant. The voltage regulation of an ac generator is defined as the rise in voltage when the load is reduced from full-rated value to zero, speed and field current remaining constant. It is normally expressed as a percentage of full-load voltage,

$$\text{voltage regulation} = \frac{V_{NL} - V_{FL}}{V_{FL}} \times 100\% \tag{6.5}$$

where V_{NL} is the no-load terminal voltage, which equals the generated EMF, and V_{FL} is the full-load terminal voltage.

The subtraction in Eq. 6.5 is done algebraically and not vectorially. Figure 6.2 shows that the percent regulation varies considerably depending on the power factor of the load. For a leading load power factor it even becomes negative, which tells us that the terminal voltage increases upon loading.

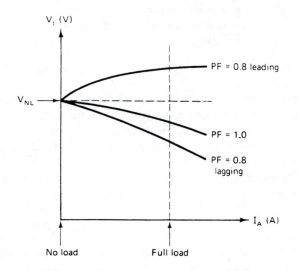

FIGURE 6.2. Terminal voltage versus armature current for different load power factors.

EXAMPLE 6.2

When the load is removed from an ac generator its terminal voltage rises from 480 V at full load to 660 V at no load. Calculate the voltage regulation.

Solution　From Eq. 6.5, the voltage regulation is

$$\frac{660 - 480}{480} \times 100 = 37.5\%$$

In general, the voltage variation is appreciable when the load is removed. It is for this reason that automatic voltage regulators are used in conjunction with synchronous generators. This inherent voltage change cannot be compensated for (as with dc generators) by the use of compounding. The voltage variation, particularly at low lagging power factors, is considerably larger than those displayed by dc machines, due to the armature reaction and armature reactance effects. Also, generators normally feed long transmission lines consisting of wires and transformers, the impedances of which introduce additional voltage drops. These combined factors act simultaneously to cause large voltage fluctuations with changing loads, which are intolerable in power distribution systems. Therefore, specially designed regulators are employed to act on the dc field excitation so that a drop in voltage is accompanied by an increase in the flux.

Generator Voltage Drops

When a load is placed on a generator, for example, a three-phase motor is connected to it, current will flow in the generator armature winding. The armature winding has an armature resistance R_A, in ohms per phase. The result is an armature voltage drop

$$V_A = I_A R_A \quad \text{V/phase} \tag{6.6}$$

where I_A is the phase current in amperes and R_A is the winding resistance in Ω/phase. As the three phases are completely symmetrical, the voltage drops will be the same in all three phases. This also applies to the current. It is for this reason that all values used are phase values, with the understanding that all phases behave the same. Calculations done as such are said to be referred to on a *per phase basis*.

In addition to the resistive voltage drop there is a reactive voltage drop due to the armature winding inductance. Its value is

$$V_X = I_A X_A \quad \text{V/phase} \tag{6.7}$$

where X_A is the armature winding reactance/phase, as determined by Eq. 6.4.

Knowing the magnitudes of these voltage drops alone is not sufficient to determine the generated voltage. The voltages must be added vectorially to the terminal voltage; thus it is essential that for a given load current the power factor is also known. Figure 6.3 readily demonstrates the dependence of E_G on the power factor. For unity and lagging power

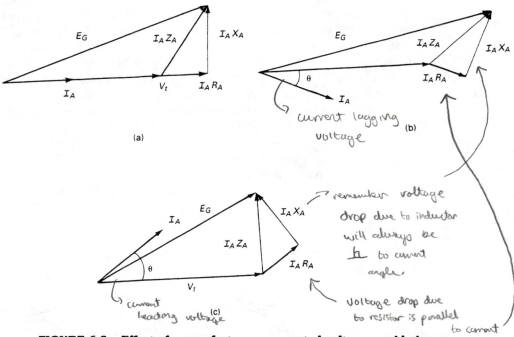

FIGURE 6.3. **Effect of power factor on generated voltage considering armature resistance and reactance only, for: (a) PF = 1.0; (b) PF lagging; (c) PF leading.**

factors, E_G is larger than V_T. For a leading power factor the opposite occurs, namely, V_T is larger than E_G. Note, however, that E_G leads V_T in phase for all power factors.

We will now consider the third factor responsible for the voltage change upon loading the generator. This is the armature reaction effect, which plays a significant role in the ultimate value of generated voltage.

Armature Reaction

If the rotor field in the generator is excited and no current flows in the armature winding (no-load condition), the flux paths for a two-pole machine are as shown in Figure 5.6. When the armature winding is carrying current (loaded condition), it in turn will set up its own field. When both fields are present simultaneously, as it would be under loading conditions, they will react with each other and a single resultant flux pattern will exist.

This is not unlike armature reaction effects as we discussed in dc machines. It is thus expected that the generated voltage in synchronous generators is affected by armature reaction as well. It is, but the nature of the resultant field is now dependent on the power factor of the load and is instrumental in establishing the resultant field pattern.

Figure 6.4 shows a portion of the field structure rotating clockwise. Assume the coils on the armature are full pitch. When the adjacent pole centers are opposite the coil sides, as in Figure 6.4a, maximum voltage is induced in the coil. The direction of the induced voltage is readily confirmed by the right-hand rule. If the current in the coil is in phase with the induced voltage (i.e., the power factor is unity), a flux is set up around the coil indicated by the arrows. The direction of this armature coil flux is seen to aid the flux of the north pole and oppose the flux of the south pole within the area of the coil span. The amounts added and subtracted just about balance each other, except that the degree of saturation in the pole shoe tips is slightly different. This is the result of the nonlinear behavior of the magnetic steel. The net result is, as we may recall from our discussion of armature reaction in Chapter 2, a net reduction of flux cutting the armature coil.

For a zero-power-factor lagging current (i.e., the current lags the induced voltage by 90°), the current in the armature coil will reach its maximum when the rotor has moved

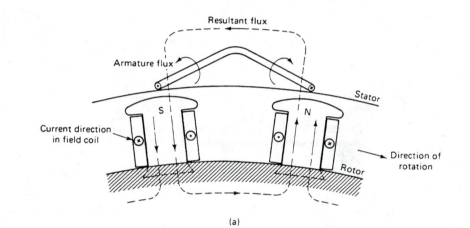

(a)

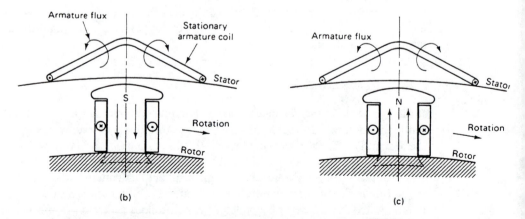

(b)   (c)

FIGURE 6.4. Armature reaction fluxes: (a) unity PF; (b) PF = 0 lagging; (c) PF = 0 leading.

through 90 electrical degrees as compared to the position shown in Figure 6.4a. This implies the south pole will center under the coil axis, as shown in Figure 6.4b. The armature coil flux is now directly opposed to the main pole flux and acts to demagnetize the poles.

For a zero-power-factor leading current, the armature coil current will have reached its maximum point 90 electrical degrees earlier as compared to the position shown in Figure 6.4a. In this situation the armature coil will center under the north-pole axis, as shown in Figure 6.4c. Here, as we can see, the armature flux is completely aiding the main pole flux and acts to magnetize the poles. In most practical situations the armature current will neither be in phase nor leading or lagging the voltage by 90 electrical degrees, but will be somewhere in between.

In general, the armature field produces a change in magnitude and distribution of the main pole field. It means the net air gap flux is modified, which translates in a lower or higher generated voltage E_G, consequently V_T. The illustrated generator characteristics in Figure 6.2 can now be appreciated. In total, there are three factors affecting the generated voltage:

1. the armature resistance voltage drop,
2. the armature inductance voltage drop, and
3. the armature reaction effect.

We will see that the armature reaction produces an effect similar to that which would be obtained by adding additional reactance to the armature circuit. It, too, can therefore be treated as an extra reactive voltage drop.

At this point in our discussion, let us represent the various quantities in a phasor diagram to obtain a qualitative idea of what is happening. Figure 6.5a applies to the zero-power-factor lagging load current (PF = 0, the load is thus purely inductive). Notice that the main pole flux ϕ_F is responsible for the no-load generated voltage $\mathbf{E}_G$, the latter being taken as the reference phasor. The EMF $\mathbf{E}_G$ lags ϕ_F by 90 electrical degrees, since an induced voltage lags behind the flux that produces it by 90°. With the generator supplying a zero-PF lagging load current $\mathbf{I}_A$, the resulting armature flux ϕ_{AR} (in phase with the current $\mathbf{I}_A$) develops a voltage $\mathbf{V}_{AR}$ in the armature winding. Again $\mathbf{V}_{AR}$ lags the flux ϕ_{AR} that induces it by 90°. It is now apparent what is happening. The armature reaction flux ϕ_{AR} reacts with the main pole field flux. Since they are oppositely directed (180° out of phase) we can obtain the resultant flux ϕ_R simply by subtracting ϕ_{AR} from ϕ_F. Normally, these two fluxes are never directly opposite, and they have to be added vectorially. This we indicate by,

$$\phi_R = \phi_F + \phi_{AR} \quad \text{Wb} \tag{6.8}$$

where ϕ_R = the resultant air gap field, Wb
  ϕ_F = the main pole field, Wb
  ϕ_{AR} = the armature reaction field, Wb

Boldface type indicates vector quantities, which implies they have magnitude and direction. Similar reasoning is now extended to the voltages generated. The armature reaction effect is responsible for the voltage $\mathbf{V}_{AR}$. It must be added to $\mathbf{E}_G$ vectorially to obtain the terminal

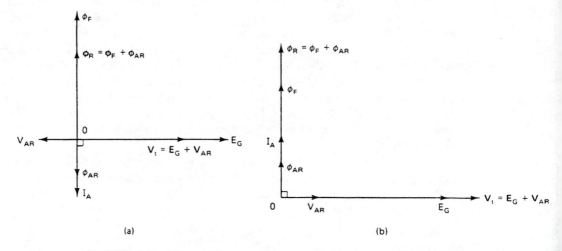

FIGURE 6.5.   Phasor diagram for zero PF load current: (a) lagging; (b) leading.

voltage. Thus

$$\mathbf{V}_t = \mathbf{E}_G + \mathbf{V}_{AR} \quad \text{V} \tag{6.9}$$

where $\mathbf{V}_t$ = the generated terminal voltage, V
$\mathbf{E}_G$ = the generated EMF, V
$\mathbf{V}_{AR}$ = the armature reaction voltage, V

As shown, the two voltages $\mathbf{E}_G$ and $\mathbf{V}_{AR}$ are also 180 electrical degrees apart, so that the vectorial addition amounts to subtracting $\mathbf{V}_{AR}$ from $\mathbf{E}_G$ to arrive at $\mathbf{V}_t$. In most practical applications they too are not 180° apart.

It is now an easy matter to extend this discussion to the zero-power-factor leading load current (i.e., purely capacitive load, again PF = 0). This results in the phasor diagram of Figure 6.5b. Applying Eq. 6.8 to the fields results as expected in a net air gap field that is larger than the main pole field, because the armature reaction field is aiding the main pole field. Similarly, the resulting terminal voltage V_t is larger than the generated voltage by an amount equal to V_{AR}. Equation 6.9 confirms this. The magnitudes of ϕ_{AR} and V_{AR} naturally depend on the load current, and the specific phase angle will be determined by the power factor of the load. This is what the next section is all about.

6.4   General Phasor Diagram

Figure 6.6 shows three phasor diagrams for the general types of loads. For the resistive load (PF = 1.0), E_G is larger than V_t. For the lagging power factor load, the change from V_t to E_G is comparatively greater than it is for unity power factor load. But when the load power factor becomes leading, V_t becomes larger than E_G. Thus when the load is removed, the generated voltage E_G now appearing at the machine terminals is smaller than it was under full load. From the construction of Figure 6.6 it is evident that the voltage drops $I_A X_{AR}$ and $I_A X_A$ are in phase. This implies that the armature reaction produces an effect similar to that of an additional reactance added in the armature circuit. The summation of the voltage drops V_{AR} and $I_A X_A$ gives the total reactive voltage drop produced in the armature circuit by the armature field. It results in

armature winding reactance

$$I_A(X_A + X_{AR}) = I_A X_S \tag{6.10}$$

↳ armature reaction reactance

where $X_S = X_A + X_{AR}$. This voltage drop is normally referred to as the synchronous reactance voltage drop and X_S is called the *synchronous reactance*. It is the value of X_S which can be determined from tests, as will be discussed in the next section.

An interesting observation with regard to Figure 6.6 should be pointed out. The angle δ indicated between $\mathbf{E}_G$ and $\mathbf{V}_T$ is called the *power angle* or *torque angle* of the machine. It varies with load and is a measure of the air gap power developed in the machine. The reader may note that the power angles are zero in Figure 6.5, because a purely inductive or capacitive load does not consume power.

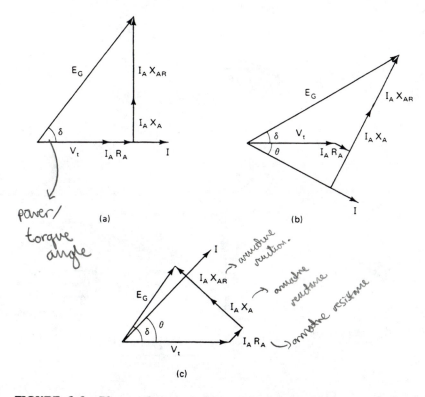

power/ torque angle

→ armature reaction.

→ armature reactance

→ armature resistance

FIGURE 6.6. Phasor diagram of a synchronous generator under three types of loading conditions: (a) unity PF; (b) PF lagging; (c) PF leading.

6.5 *Voltage Regulation from Equivalent Circuit*

It is clear that when dealing with the synchronous reactance $X_S = (X_A + X_{AR})$, the distinction between armature reaction and armature reactance voltage drops need not to be made. Therefore, the phasor diagram reduces to that shown in Figure 6.7a. With reference to this figure, using $\mathbf{V}_t$ as the reference phasor, $\mathbf{E}_G$ can be solved for, since

$$\mathbf{E}_G = \mathbf{V}_t + \mathbf{I}_A\mathbf{Z}_S \quad \text{V} \tag{6.11}$$

where Z_S is the armature impedance/phase $= R_A + jX_S$ ohms and the other parameters as defined before.

The reader may recall from circuit theory that the phasor diagram in Figure 6.7a is applicable to the equivalent circuit represented in Figure 6.7b. This suggests that the synchronous generator on a per phase basis can be represented by an R–L circuit. The resistance represents the armature winding resistance R_A, and the reactance represents the synchronous reactance X_S, both phase quantities. The power factor angle θ in the circuit is dictated by the load connected to the generator terminals.

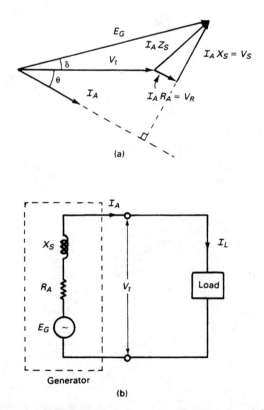

FIGURE 6.7.   Simplified equivalent ac circuit (per phase) for synchronous generator: (a) phasor diagram; (b) electric circuit model.

The power consumed by the load is

$$P_L = V_t I_A \cos\theta \quad \text{W/phase} \tag{6.12}$$

The machine is a three-phase synchronous generator, thus

$$P_L = 3V_t I_A \cos\theta \quad \text{W} \tag{6.13}$$

In general we specify power in terms of line quantities, that is, line current I_L and line (between machine terminals) voltage V_L. To convert Eq. 6.13 in those terms, we have to know how the generator winding is connected. For a Y-connected armature winding, which is normally done to bring out the grounded neutral, $I_A = I_L$ and $V_t = V_L/\sqrt{3}$.

Making these substitutions in Eq. 6.13 gives

$$P_L = 3\frac{V_L}{\sqrt{3}} I_L \cos\theta = \sqrt{3}V_L I_L \cos\theta \tag{6.14}$$

which is the general expression for three-phase power. Equation 6.14 gives the power received or delivered in a three-phase system and is independent on either armature winding or load connection.

To determine the voltage regulation, we need the generated voltage E_G. This can be obtained from Eq. 6.11. Write I_A and Z_S in phasor form as

$$\mathbf{I}_A = I_A \angle -\theta \tag{6.15}$$

and

$$\mathbf{Z}_S = R_A + jX_S = Z_S \angle \beta \tag{6.16}$$

in which $\mathbf{V}_t$ is taken as the reference phasor (i.e., $\mathbf{V}_t = V_t \angle 0°$), and $\beta = tan^{-1}(X_S/R_A)$. Thus for inductive loads, which are characterized by a lagging power factor, we obtain, by substituting Eqs. 6.15 and 6.16 into Eq. 6.11,

$$\mathbf{E}_G = \mathbf{V}_t \angle 0 + I_A \angle -\theta \times Z_S \angle \beta \tag{6.17}$$

To solve Eq. 6.17 for E_G, the equivalent circuit parameters for the generator as well as the load details must be known. In Section 6.6 we discuss the tests that are necessary to determine R_A and X_S.

As is apparent, a phasor diagram is an extremely useful tool of solving machine problems. It provides a visual aid of how various parameters relate to each other. The influence of specific quantities on others can readily be appreciated.

To complete the discussion, assume a capacitive load is put on the generator. This results in an armature current leading the terminal voltage (see Figure 6.8). Taking the terminal voltage as a reference, there results

$$\mathbf{E}_G = \mathbf{V}_t \angle 0 + I_A \angle \theta \times Z_S \angle \beta \tag{6.18}$$

When the load is purely resistive, the power factor is unity, and $\theta = 0$, which reduces the expression to

$$\mathbf{E}_G = \mathbf{V}_t + I_A Z_S \angle \beta \tag{6.19}$$

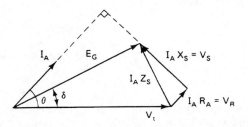

FIGURE 6.8. Phasor diagram for a synchronous generator supplying a capacitive load (parameters on a per phase basis).

In practical generators the value of R_A is normally much smaller than X_S, particularly for the larger size machines. This implies that $Z_S \approx jX_S$ can be used to simplify a problem.

EXAMPLE 6.3

A 250-kVA 660-V 60-Hz three-phase synchronous generator is Y-connected as shown in Figure 6.9. The armature resistance is 0.10 Ω/phase and the synchronous reactance is 1.40 Ω/phase. Determine the voltage regulation for a load having a power factor of 0.866 lagging.

Solution   As indicated in the text, all calculations will be made on a per phase basis because the voltage change is the same as that obtained between line terminals. At rated load,

$$I_A = \frac{kVA \times 1000}{\sqrt{3} \times V_L} = \frac{250,000}{\sqrt{3} \times 660} = 219 \text{ A } \underline{/-30°}$$

determined by P.F.

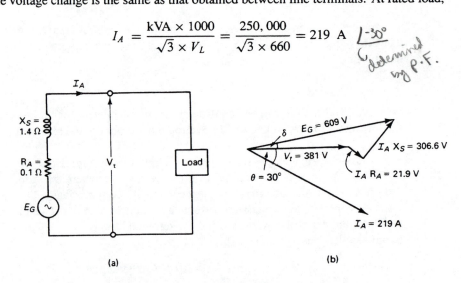

(a)

(b)

FIGURE 6.9. (a) Equivalent circuit diagram for Example 6.3 and (b) its corresponding phasor diagram, all per phase quantities.

which is the armature phase current.

(handwritten: ↱ observe how V_t treated as reference)

$$V_t = \frac{V_{L-L}}{\sqrt{3}} = 381 \text{ V/phase}$$

Then at 0.866 lagging power factor (i.e. $\theta = 30°$), and

$$Z_S = \sqrt{0.01^2 + 1.4^2} \; \angle tan^{-1}(1.4/0.10) = 1.404\angle 85.9°$$

(handwritten: ↳ NOT NECESSARY! ↳ WHAT DO IT?)

there results from Eq. 6.17,

$$
\begin{aligned}
E_G &= 381\angle 0 + 219\angle -30 \times 1.404\angle 85.9 \\
&= 381 + 307.5\angle 55.9 = 609.2\angle 24.7°
\end{aligned}
$$

Hence the voltage regulation is

$$\frac{609.2 - 381}{381} \times 100 = 59.9\%$$

Note that E_G leads the reference phasor V_t by 24.7°. This is the power factor angle δ.

6.6 Synchronous Impedance and Reactance

In this section we discuss how the equivalent circuit parameters are obtained. From Example 6.3 it can be seen that for any given load the voltage regulation can be determined, provided we have the machine's electric circuit model. Voltage regulation is, of course, of prime concern, since the line voltage must be kept constant within reasonable limits when the load on the generator changes.

To determine the voltage regulation on a large machine experimentally is practically not feasible. It would require exactly full-load conditions while maintaining the terminal voltage, frequency, and predetermined power factor of the load.

In actual practice the regulation is calculated from data obtained from a series of relatively simple tests in order to arrive at the equivalent circuit parameters. The quantities to be determined are (1) the synchronous reactance X_s, and (2) the armature resistance R_A, both on a per phase basis.

Several methods are in existence to determine the synchronous reactance, only one will be discussed, known as the synchronous impedance method. To obtain the necessary data, three simple tests are required:

1. The open-circuit test.
2. The short-circuit test.
3. The armature resistance test.

Open-Circuit Test

This test is performed at synchronous speed with the armature circuit open-circuited, as shown in Figure 6.10a. Provisions must be made to adjust the field current I_f, so that when starting at zero, it may be raised until the line-to-line output voltage is somewhat above rated voltage. Recording I_f and V_{oc} (open-circuit terminal voltage) for a sufficient number of steps, the *open-circuit saturation curve* may be plotted (see Figure 6.11). In plotting this curve, the open-circuit line-to-line voltage V_{oc} is divided by $\sqrt{3}$, to represent the per phase value, which is then plotted as a function of I_f.

Short-Circuit Test

In reference to Figure 6.10b, the generator terminals are short circuited with an ammeter in one of the lines. The field current is reduced to zero and the generator operated at rated speed. The field current is then gradually increased to a safe maximum armature current obtainable, possibly at twice rated value. Care must be taken not to allow this armature current to remain at high values for a long period of time.

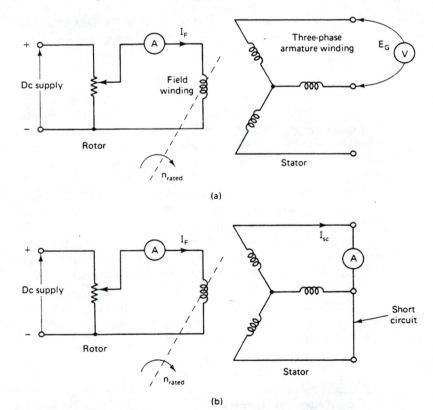

FIGURE 6.10.   Equivalent circuits for synchronous generator under test: (a) open-circuit test; (b) short-circuit test (Y-connected armature winding is assumed).

During this test the values of I_f and the short-circuit current I_{sc} are recorded. Plotting I_{sc} versus I_f gives the *short-circuit characteristic*, which is also illustrated in Figure 6.11. Referring now to Figure 6.10b, we see that with the load terminals short-circuited, the generated voltage is equal to the synchronous impedance drop. By maintaining the field current constant and removing the short circuit, the open-circuit voltage then produced is obviously E_{Goc}. Thus with I_f known, the open-circuit characteristic can be used as shown in Figure 6.11 to obtain E_{Goc}. Similarly, at this value of field current the corresponding short-circuit current that resulted can be found. The synchronous impedance is then the ratio of these two quantities, namely,

$$Z_s = \left. \frac{E_{Goc}}{I_{sc}} \right|_{I_f = constant} \quad \Omega \tag{6.20}$$

The value of synchronous impedance that is obtained can be seen to depend on the value of I_f chosen. Therefore, we will select the value of I_f that results in rated armature current, which results in a synchronous impedance,

$$Z_s = \frac{E_{Goc}}{I_{Asc}} \tag{6.21}$$

where I_{Asc} is the rated armature current from the short-circuit characteristic, which at the same field current I_f gives the open circuit terminal voltage of the machine E_{Goc}, on the open-circuit characteristic.

Resistance Test

Having determined the synchronous impedance, the synchronous reactance X_s can be obtained, knowing the armature resistance. This can, of course, be determined from a simple

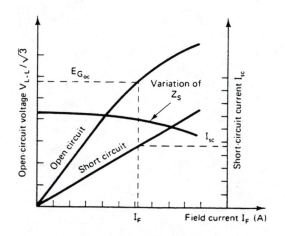

FIGURE 6.11.   Open-circuit and short-circuit characteristic curves for a synchronous generator, on a per phase basis.

resistance test. For this, assume that the generator is Y-connected. This assumption does not in any way affect the final result, since an identical voltage regulation is obtained if a Δ connection is assumed. With the field circuit open, the dc resistance is measured between any two of the output terminals. Since during this test two phases are in series, the armature resistance per phase is half the measured value. In practice, this resistance value is multiplied by a factor to arrive at the effective ac resistance, R_{Aeff}. This factor depends on the slot size and shape, the size of the armature conductors, and the particular winding construction. It ranges on practical machines between 1.2 and 1.5 depending on the size of the machine. A typical value to use in the calculations would be 1.25. Then

$$X_s = \sqrt{Z_s^2 - R_{Aeff}^2} \tag{6.22}$$

EXAMPLE 6.4

The 500-kVA 2300-V three-phase synchronous generator shown in Figure 6.12 was tested according to the test procedure outlined to determine its voltage regulation at a full-load power factor of 0.866 lagging. The data obtained are:

> Dc resistance test: $V_{L-L} = 8$ V, $I_L = 10$ A
> Open-circuit test: $I_f = 25$ A, $V_{L-L} = 1408$ V
> Short-circuit test: $I_f = 25$ A, $I_{sc} = 126$ A = rated full-load current

Assume that the generator is Y-connected.

Solution

$$V_{oc} = \frac{V_{L-L}}{\sqrt{3}} = \frac{1408}{\sqrt{3}} = 812.9 \text{ V/phase}$$

$$Z_s = \frac{V_{oc}}{I_{Asc}} = \frac{812.9}{126} = 6.45 \ \Omega$$

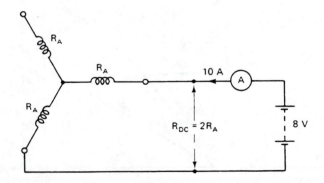

FIGURE 6.12. DC resistance measurement for Example 6.4.

$$R_{A(eff)} = 1.25 \times \frac{8\ V}{2 \times 10\ A} = 0.50\ \Omega$$

where the factor 1.25 represents the correction necessary to the dc resistance to arrive at the effective ac resistance. Since during the resistance measurement two phases are in series, the measured value must be divided by 2.

$$X_s = \sqrt{Z_s^2 - R_A^2} = \sqrt{6.45^2 - 0.5^2} = 6.43\ \Omega$$

Therefore,

$$\beta = tan^{-1}(6.43/0.50) = 85.6°$$

and

$$V_t = \frac{V_{L-L}}{\sqrt{3}} = 1328\ \text{V/phase at full load}$$

The generated voltage per phase from Eq. 6.17 using V_t as the reference phasor, and $\theta = cos^{-1}0.866 = 30°$ is then

$$
\begin{aligned}
E_G &= 1328\angle 0 + 126\angle -30 \times 6.45\angle 85.6 \\
&= 1787.1 + j670.6 = 1909\angle 20.6°\ \text{V}
\end{aligned}
$$

The voltage regulation is then

$$\frac{1909 - 1328}{1328} \times 100 = 43.8\%$$

6.7 Developed Power

In electromechanical energy conversion devices (and the synchronous generator is no exception), the expression of the developed power is important. It should be stressed that we are converting the mechanical input power from the prime mover into electrical output power. It was shown in Chapter 2 that once the developed torque T is calculated, the developed power P_d can be found from

$$P_d = \omega T \quad \text{W} \tag{6.23}$$

where ω is the mechanical speed of rotation in radians per second.

It is often useful, however, to have an equation that expresses this power in terms of machine parameters such as voltage and phase angle. A better insight regarding machine performance may then be obtained.

To do so, let us refer to Figure 6.13. It shows the synchronous machine represented by a reactance X_s and the corresponding phasor diagram for this circuit. E_G is the internal generated voltage and V_t the terminal voltage, both on a per phase basis. We have neglected the armature resistance. This is acceptable, since the effect of armature resistance is very

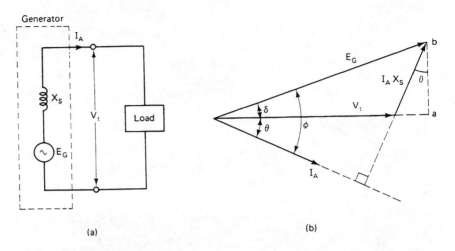

FIGURE 6.13. Synchronous generator: (a) circuit diagram; (b) phasor diagram (armature resistance is assumed negligible).

small in practical generators. We can write the expression for the developed power in one of two ways:

$$P_d = 3V_t I_A \cos\theta \quad \text{W} \tag{6.24}$$

or

$$P_d = 3E_G I_A \cos\phi \quad \text{W} \tag{6.25}$$

where $\phi = \delta + \theta$ electrical degrees.

Equation 6.25 follows from Figure 6.13b upon noting that the projection of E_G on the I_A phasor is identical with the projection of V_t on I_A. Similarly, we can see that the quantity ab in Fig. 6.13b can also be expressed in two ways: namely,

$$ab = E_G \sin\delta = I_A X_s \cos\theta \tag{6.26}$$

From Eq. 6.26 it follows that

$$I_A \cos\theta = \frac{E_G}{X_s} \sin\delta \tag{6.27}$$

If we now substitute the result of Eq. 6.27 into Eq. 6.24, the desired result will be obtained:

$$P_d = 3\frac{V_t E_G}{X_s} \sin\delta \quad \text{W} \tag{6.28}$$

Equation 6.28 applies only under the assumption that R_A is negligible. The factor of 3 refers to the three phases. Because this power is seen to depend on the angle δ, it is called the *power* or *torque angle*. Thus Eq. 6.28 gives the developed power in a synchronous generator where the armature resistance is negligible. As Eq. 6.28 indicates, no power

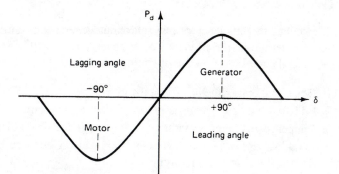

FIGURE 6.14. Graphical representation of Eq. (6.28), P_d versus δ.

is generated when the power angle δ is zero. Furthermore, the equation reveals that the developed power is a sinusoidal function of the power angle. Its maximum occurs when $\sin \delta = 1.0$ or $\delta = 90°$. Figure 6.14 graphically displays the variation of P_d versus positive and negative values of δ. Positive angles refer to generator action. Negative angles refer to motor action.

EXAMPLE 6.5

A three-phase synchronous generator delivers power to an electric distribution system at 13.8 kV. The generator reactance is 8 Ω/phase. Its resistance is negligible. The magnitude of the generated EMF is 20% higher than the machine terminal voltage. When the machine delivers 12 MW to the system, determine the power angle δ at which the machine is operating.

Solution The terminal voltage

$$V_t = \frac{13,800}{\sqrt{3}} = 7967 \text{ V/phase}$$

Since the generated EMF is 20% larger than this,

$$V_t = 7967 \times 1.2 = 9560 \text{ V/phase}$$

The delivered power = 12/3 = 4 MW/phase. From Eq. 6.28 we obtain

$$4 = \frac{7.967 \times 9.56}{8} \sin \delta$$

Therefore,

$$\sin \delta = 0.42 \quad \text{and} \quad \delta = 24.8°$$

6.8   *Losses and Efficiency*

To calculate the efficiency of a synchronous generator, a similar procedure is followed to that of determining the efficiency of dc generators. It will be recalled that it is necessary to establish the total losses when operating under load. For generators this includes:

1. Rotational losses such as friction and windage.
2. Eddy current and hysteresis losses in the magnetic circuit.
3. Copper losses in the armature winding and in the field coils.
4. Load loss due to armature leakage flux, causing eddy current and hysteresis losses in the armature surrounding iron.

With regard to the losses the following comments may be made:

1. The rotational losses, which include friction and windage losses, are constant, since the speed of a synchronous generator is constant. They may be determined from the no-load test.
2. The core loss includes eddy current and hysteresis losses as a result of normal flux density changes. It can be determined by measuring the power input to an auxiliary motor, used to drive the generator at no load, with and without the field excited. The difference in power measured constitutes this loss.
3. The armature and field copper losses are obtained as $I_A^2 R_A$ and $V_f I_f$. Since per phase quantities are dealt with, the armature copper loss for the generator must be multiplied by the number of phases. The field winding loss is as a result of the excitation current flowing through the resistance of the field winding.
4. Load loss or stray losses result from eddy currents in the armature conductors and increased core losses due to distorted magnetic fields. Although it is possible to separate this loss by tests, in calculating the efficiency it may be accounted for by taking the effective armature resistance rather the dc resistance.

After all the foregoing losses have been determined, the efficiency η is calculated as

$$\eta = \frac{\text{kVA} \times \text{PF}}{\text{kVA} \times \text{PF} + (\text{total losses})} \times 100\% \qquad (6.29)$$

where η = efficiency, percent
       kVA = load on the generator
        PF = power factor of the load

The quantity (kVA × PF) is, of course, the real power delivered to the load (in kW) by the synchronous generator. Thus Eq. 6.29 could in general be stated as

$$\eta = \frac{P_{out}}{P_{in}} \times 100 = \frac{P_{out}}{P_{out} + P_{losses}} \times 100\% \qquad (6.30)$$

The input power $P_{in} = P_{out} + P_{losses}$ is the power required from the prime mover to drive the loaded generator. To calculate the load power P_{out}, the reader may also use the expressions given by Eqs. 6.13 and 6.14. It should be noted that in determining the efficiency of a practical generator, the power supplied to the field circuit is not always included.

As we indicated previously, in large machines the power supplied to the field by the excitation system is large. However, when we compare this power input with that supplied by the prime mover, we quickly realize that in the overall scheme the field power is rather small. As a matter of fact, it represents only a fraction of a percent.

To give some practical figures on this, a 1000-MVA synchronous generator (which is a very large machine, even by today's standards) would require a 6- to 7-MW field circuit supply. This amounts to about 0.6 to 0.7% of the prime-mover input power. This is no small task. Realizing that the generated voltage is on the order of 24 kV, the field supply at the 500-V level, the currents involved are enormous. This, of course, calls for forced cooling in all parts of the machine to keep the generated heat created by the various losses to acceptable levels. On the plus side, with forced cooling the allowable current density (current per unit area) in the copper can be at least doubled. This keeps conductor sizes "reasonable." The overall result is a greatly reduced physical size of the machine when forced cooling is implemented. It is for this reason, then, that the field circuit loss is not always considered when calculating efficiencies. Its inclusion would barely alter the results.

EXAMPLE 6.6

A 2000-kVA 2300-V three-phase synchronous generator has a dc armature winding resistance of 0.032 Ω between terminals. The field takes 32 A from a 250-V dc supply. Friction and windage loss is 12.8 kW, and core loss including stray load losses is 10.6 kW. Calculate the efficiency of the generator at full load and a power factor of 0.80 lagging. Assume a Y-connected generator. The effective armature resistance may be taken as 1.25 times its dc value.

Solution   The armature current

$$I_A = \frac{\text{kVA} \times 1000}{\sqrt{3} \times V_{L-L}} = \frac{2,000,000}{\sqrt{3} \times 2300} = 503 \text{ A}$$

$$R_A \text{ (per phase)} = \frac{0.032}{2} \times 1.25 = 0.020 \text{ } \Omega$$

Losses	kW
Friction and windage	12.8
Core loss and stray loss	10.6
Armature winding	
$3 \times 503^2 \times 0.020$	15.18
Field winding $(250 \times 32)/1000$	8.0
Total loss	46.58

The efficiency

$$\eta = \frac{P_{out}}{P_{out} + P_{losses}} \times 100$$

$$= \frac{2000 \times 0.80}{2000 \times 0.8 + 46.58} = 97.2\%$$

It can be shown that the maximum efficiency occurs at a load where the constant losses equal the variable losses. Those that are considered constant include friction and windage, core loss, and field copper loss. The armature copper losses are variable since they depend on the armature current. For most electrical machines (and the synchronous generator is no exception) the efficiency improves rapidly from a low value to a maximum with increasing load. The point at which the efficiency reaches a maximum can be controlled at the design stage. Generally, it is not selected at 100% loading conditions. Machines seldom run at full load continuously; this applies to transformers and motors alike.

ASSIGNMENT PROBLEMS (CIRCLED.

PROBLEMS

6.1 A four-pole synchronous generator is driven at 1800 r/min. The maximum field-pole flux is 0.032 Wb. Each phase winding has 280 turns. Calculate the generated EMF per phase.

6.2 If the generator of Problem 6.1 is rated at 800 kVA and is Y-connected, what is the rated current?

6.3 If the generator of Problem 6.1 is Δ-connected and delivers 600 kW at a power factor of 0.866, calculate the line current.

6.4 An eight-pole three-phase armature has a total of 504 conductors. The flux per pole is 0.0128 Wb and is sinusoidal. If the frequency is 60 Hz, find the generated voltage/phase.

6.5 The maximum flux per pole of an eight-pole generator is 0.016 Wb, the rated frequency is 60 Hz. There are a total of 1728 conductors in the slots. What is the open-circuit line voltage if the armature is Δ-connected and each phase winding consists of two parallel paths?

6.6 A 250-kVA 660-V 60-Hz three-phase synchronous generator is Y-connected. The effective armature resistance is 0.2 Ω/phase and the synchronous reactance is 1.4 Ω/phase. At full load and unity power factor, calculate the voltage regulation.

6.7 Calculate the voltage regulation of the generator in Problem 6.6 at full load and

    a. 0.866 lagging power factor
    b. 0.70 leading power factor

6.8 A 1000-kVA 4600-V three-phase 60-Hz Y-connected ac generator has a no-load voltage of 8350 V. The generator is operated at rated volt-amperes and 0.75 power factor lagging and rated voltage. Calculate

    a. the synchronous reactance (neglect armature resistance),
    b. the voltage regulation,
    c. the torque angle,
    d. the developed or air gap power, and

e. the new voltage and kVA rating if the armature winding is reconnected in delta.

6.9 A 100-kVA synchronous generator has a total loss of 5.5 kW when it delivers rated kVA at a power factor of 0.8. Calculate the percent efficiency.

6.10 A 25-kVA 220-V three-phase generator delivers rated kVA at a power factor of 0.8 lagging. The effective ac resistance between Y-connected armature winding terminals is 0.20 Ω. The synchronous reactance is 0.6 Ω/phase. The field winding current is 9.3 A at 115 V dc. The friction and windage loss is 460 W and the magnetic (iron) core loss is 610 W. Calculate

a. the voltage regulation, and
b. the efficiency at full load.

6.11 One of the three-phase waterwheel generators at Grand Coulee Dam, Washington, is rated at 108,000 kVA, PF unity, 13.8 kV, Y connected, 60 Hz, 120 r/min. Determine

a. the number of poles,
b. the kW rating,
c. the current rating,
d. the input at rated kW load if the efficiency is 97% (excluding the field loss), and
e. the prime-mover torque applied to the generator shaft.

6.12 If a field current of 10 A in a synchronous generator results in a current of 150 A on short circuit and a terminal of 720 V on open circuit, determine the internal voltage drop when the generator delivers a load current of 60 A. Armature resistance may be neglected.

6.13 A three-phase synchronous generator operates onto a 13.8-kV power grid. The synchronous reactance is 5 Ω/phase. It is delivering 12 MW and 6 MVar (inductive) to the system. Determine

a. the power angle,
b. the phase angle, and
c. the generated EMF E_G

6.14 A three-phase 750-kVA 2300-V synchronous generator is short-circuited and is operated at rated speed to produce rated current. With the short circuit removed and the same excitation, the voltage between stator terminals is 1320 V. The armature resistance may be neglected. Determine the percent regulation of the generator at a power factor 0.8 leading.

6.15 Repeat Problem 6.14 for a power factor 0.8 lagging.

6.16 A 2400-V 2000-kVA three-phase delta-connected synchronous generator has a synchronous reactance of 1.5 Ω/phase, its armature resistance may be neglected. The excitation of the generator is adjusted to generate rated voltage at no load. Calculate the generator terminal voltage when delivering rated current at 0.80 power factor lagging.

7

Transformers

7.1 Introduction

In Chapter 5 the need for high-voltage transmission was discussed for the case in which electrical power is to be provided at considerable distances from a generating station. At some point this high voltage must be reduced, because ultimately it must supply a load. The transformer makes it possible for various parts of a power system to operate at different voltage levels. In this chapter we discuss power transformer principles and applications.

7.2 Two-Winding Transformers

A transformer in its simplest form consists of two stationary coils coupled by a mutual flux (see Figure 7.1). The coils are said to be mutually coupled because they both link a common flux. In power applications, laminated steel core transformers (to which this book is restricted) are used. Transformers are efficient and reliable devices used principally to change voltage levels for economic reasons. Transformers are efficient because the rotational losses normally associated with rotating machines are absent, so relatively little power is lost when transforming power from one voltage level to another. Typical efficiencies are in the range 92 to 99%, the higher values applying to the larger power transformers.

As shown in Figure 7.1, the current flowing in the coil connected to the ac source is called the *primary winding* or simply the *primary*. It sets up the flux ϕ in the core, which varies periodically both in magnitude and direction. This flux links the second coil, called the *secondary winding* or simply the *secondary*. The flux is changing; therefore, it induces

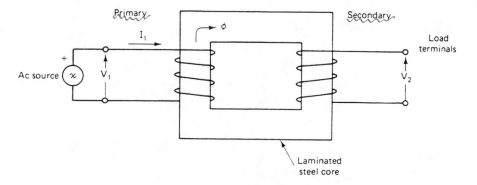

FIGURE 7.1.   Two-winding transformer.

a voltage in the secondary by electromagnetic induction in accordance with Faraday's law. Thus the primary receives its power from the source while the secondary supplies this power to a load. This action is known as *transformer action*. Before we examine this concept in detail, let us briefly look at some transformer constructional aspects.

7.3   Transformer Construction

There are basically two types of iron-core construction. They differ in the way the core is constructed to accommodate the windings. Figure 7.2a shows one type, already encountered in Figure 7.1. It consists of two separate coils, one on each of the two opposite legs of a rectangular core. Normally, this is not a desirable design. Its disadvantage is the large leakage fluxes associated with it, which results in poor voltage regulation. In reality, to achieve tighter magnetic coupling between the two coils (windings), half of each coil would be on each leg. Another construction which ensures that most of the flux set up by the primary will link the secondary is the construction shown in Figure 7.2b. This is called the *shell-type* construction. Here, the two windings are wound concentrically. The higher voltage winding is wound on top of the lower voltage winding. The low-voltage winding is then located closer to the steel, which is preferable from an electrical insulating point of view. From an electrical viewpoint there is not much difference between the two constructions.

Cores may be built up of laminations cut from alloyed sheet steel. Most laminating materials have an aproximate alloy content of 3% silicon and 97% iron. The silicon content reduces the magnetizing losses, in particular that part due to hysteresis loss. It makes the material brittle, which causes manufacturing problems, specifically in stamping operations (punching out of laminations). Therefore, there is a practical limit on the silicon content. Most laminated materials are cold-rolled and often specially annealed to orient the "grain" or iron crystals. This provides a very high permeability and low hysteresis to flux in the direction of rolling. Transformer laminations are usually from 0.01 to 0.025 in. thick for

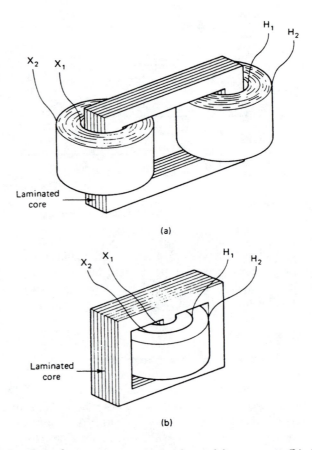

FIGURE 7.2. **Transformer core constructions: (a) core type; (b) shell type.**

60-Hz operation. They are coated on one side in one of several ways; for example, by a thin layer of varnish or paper, to insulate the laminations from each other.

Coils are prewound and the core design must be such that it permits placing the coil on the core. Of course, a core must then be made in at least two sections. The laminations for the core-type transformer of Figure 7.2a may be made of ⊔- and I-shaped laminations, as shown in Figure 7.3a. The core for the shell-type transformer of Figure 7.2b is normally made of E- and I-shaped laminations, as shown in Figure 7.3b. By stacking the lamination in alternate layers a continuous air gap is avoided, which would result in a large excitation current.

Many other popular types of core constructions are in existence, depending on manufacturer preferences. Any of the schemes involves the cost of material handling, which is an important fraction of the total cost of a transformer. As a rule, the number of butt joints are to be limited. The joints are tightly made and laminations interleaved to minimize the reluctance of the magnetic circuit. The whole core is built up or stacked to the proper

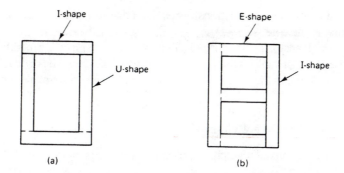

FIGURE 7.3. Core construction built of separate laminations.

dimension, and the laminations are bolted together firmly. As can be appreciated, core legs of square or rectangular cross section are the results if we stack the laminations to the required core cross section. This permits coils to be fitted on them with either square, rectangular, or circular spools or forms.

As the transformer size increases, a point is reached where this construction is unsatisfactory. Therefore, in the larger transformers, a stepped-core arrangement is used to minimize the use of copper and reduce copper loss (see Figure 7.4). This construction guarantees that each length of copper conductor embraces the maximum cross-sectional area of steel. In practical transformers, the primary and secondary windings have two or more coils per leg. They may be arranged in series or parallel, thereby providing several possible voltages.

Due to the cyclic magnetic forces and forces that exist between parallel conductors carrying current, it is necessary to clamp laminations and impregnate the coils. Insufficient clamping usually results in *hum*, giving rise to objectional audible noise. However, most hum generated by transformers is due to magnetostriction, which is the continued expansion

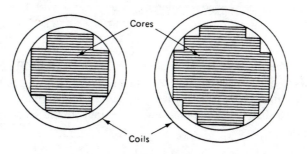

FIGURE 7.4. Stepped transformer cores.

and contraction of the transformer steel due to the increasing and decreasing (alternating) flux in the core.

Transformers are usually air cooled, even if placed in metal cases. Larger transformers are placed in tanks with special transformer oil. The oil has a dual function; it insulates while providing cooling. No matter what size, all transformers operate on the same principle.

7.4 Transformer Principles

When a sinusoidal voltage V_p is applied to the primary with the secondary open-circuited, as shown in Figure 7.5a, there will be no energy transfer. The impressed voltage causes a small current I_0 to flow in the primary winding. This no-load current has two functions: (1) it produces the magnetic flux ϕ_c in the core, which varies sinusoidally between zero and $\pm\phi_m$, where ϕ_m is the maximum value of the core flux; and (2) it provides a component to account for the hysteresis and eddy current losses in the core. These combined losses are normally referred to as the *core losses.*

The no-load current I_0 is usually a few percent of the rated or full-load current of the transformer (about 2 to 5%). Since at no load the primary winding acts as a large reactance due to the iron core, the no-load current will lag the primary voltage by nearly 90°. Figure 7.5b illustrates this relationship, where θ_0 is the no-load power factor angle. It is readily seen that the current component $I_m = I_0 \sin\theta_0$, called the *magnetizing current,* is 90° in phase behind the primary voltage V_p. It is this component that sets up the flux in the core; ϕ_c is therefore in phase with I_m.

The second component, $I_c = I_0 \cos\theta_0$, is in phase with the primary voltage. It is the current component that supplies the core losses. The phasor sum of these two components represent the no-load current, or

$$\mathbf{I}_0 = \mathbf{I}_m + \mathbf{I}_c \tag{7.1}$$

It should be noted that the no-load current is distorted and nonsinusoidal, as shown in Figure 7.6. This is the result of the nonlinear behavior of the core material.

If it is assumed that there are no other losses in the transformer, the induced voltage in the primary, E_p, and that in the secondary, E_s, can also be shown. Since the magnetic flux set up by the primary winding links the secondary winding, there will be an induced EMF E_s, in the secondary winding in accordance with Lenz's law, namely,

$$E_s = -N\frac{d\phi}{dt} \tag{7.2}$$

where the negative sign indicates that the induced EMF opposes something. It actually tends to oppose the increase or decrease in flux that is going on in the core. This will be evident in the next section when we load the transformer.

To show the magnitude and phase angle of the secondary voltage with respect to the flux, let

$\phi_m = \text{max flux}$

$$\phi_c = \phi_m \sin\omega t \tag{7.3}$$

be the flux in the core set up by the primary voltage. Then substituting this expression into Eq. 7.2 and carrying out the differentiation, gives

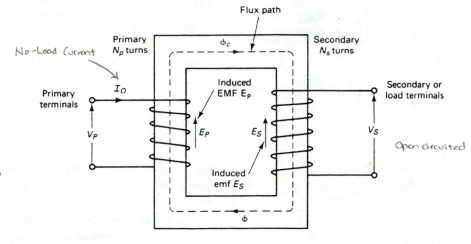

No-Load Current

Open circuited

(a)

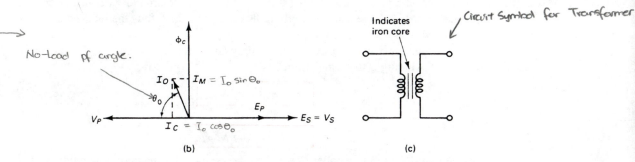

No-load pf angle.

$I_M = I_o \sin\Theta_o$

$I_C = I_o \cos\Theta_o$

(b)

Indicates
iron core

Circuit Symbol for Transformer

(c)

**FIGURE 7.5.   Transformer with open-circuited secondary;
(a) schematic; (b) phasor diagram; (c) transformer symbol.**

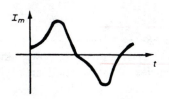

**FIGURE 7.6.   No-load current waveform
of a transformer.**

$$E_s = -N\frac{d}{dt}(\phi_m \sin\omega t)$$

$$= -N\omega\,\phi_m \cos\omega t$$

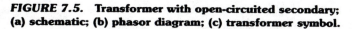

$$= E_{max}\,\sin\left(\omega t - \frac{\pi}{2}\right) \tag{7.4}$$

which indicates that E_s lags ϕ by 90°.

Furthermore, the magnitude of this voltage is

$$E_{s\,max} = N\,\phi_m\,\omega = \sqrt{2}\,E_s \qquad (7.5)$$

where E_s is the rms value of the secondary induced voltage. Thus we see that the expression for E_s, namely

$$E_s = N\,\phi_m\,\frac{\omega}{\sqrt{2}} = 4.44\,N\phi_m\,f \qquad (7.6)$$

$\omega = 2\pi f$

$\dfrac{2\pi}{\sqrt{2}}$

is in agreement with Eq. 6.3.

This same flux also links the primary itself, inducing in it an EMF, E_p. Therefore, E_p and E_s are 180° out of phase with the applied voltage. Since no current flows in the secondary winding, $E_s = V_s$. Figure 7.5c shows the circuit symbol for the transformer. The two lines between the windings identify the presence of the iron core. Note that E_p and E_s in Fig. 7.5b do not have the same magnitude. This implies an unequal turns ratio as we will see shortly. The no-load primary current I_o is small, a few percent of full-load current. Thus the voltage drop in the primary winding is small and V_p is nearly equal to E_p. The primary voltage and the resulting flux are sinusoidal, thus the induced quantities E_p and E_s vary as a sine function.

Since the same flux links with the primary and secondary windings, the voltage per turn in each winding is the same. Hence

$$E_p = 4.44\,f N_p \phi_m \qquad (7.7)$$

and

$$E_s = 4.44\,f N_s \phi_m \qquad (7.8)$$

where N_p and N_s are the number of turns on the primary and secondary winding, respectively. The ratio of primary to secondary induced voltage is called the *transformation* ratio or *turns ratio*. Denoting this ratio by a, it is seen that

$$a = \frac{E_p}{E_s} = \frac{N_p}{N_s} \qquad (7.9)$$

which is obtained by dividing Eqs. 7.7 and 7.8.

Assume the output power of a transformer equals its input power—not a bad assumption in practice considering the high efficiencies. What we really are saying is that we are dealing with an ideal transformer, that is, it has no losses. Thus

$$P_{in} = P_{out}$$

or

$$V_p\,I_p \times \text{primary PF} = V_s\,I_s \times \text{secondary PF}$$

where PF is the power factor. For the above-stated assumption it means that the power factor on primary and secondary sides are equal, therefore,

$$V_p\,I_p = V_s\,I_s$$

from which is obtained

$$\frac{V_p}{V_s} = \frac{I_s}{I_p} = \frac{E_p}{E_s} = a \qquad (7.10)$$

Equation 7.10 shows that the terminal voltage ratio equals the turns ratio. The primary and secondary currents, on the other hand, are inversely related to the turns ratio. The turns ratio gives a measure of how much the secondary voltage is raised or lowered in relation to the primary voltage.

EXAMPLE 7.1

A 100-kVA 2400/240-V transformer has 60 turns on the secondary winding. Calculate

a. the approximate value of primary and secondary currents,

b. the number of primary turns, and

c. the maximum flux ϕ_m in the core.

Solution

a.

$$S = V_p I_p$$

$$I_p \text{ (full load)} = \frac{\text{kVA} \times 1000}{V_p} = \frac{100,000}{2400} = 41.7 \text{ A}$$

and

$$I_s = \frac{100,000}{240} = 417 \text{ A}$$

[margin note: $V_p = 2400$, $V_s = 240$ — can use b/c finding approximate*]*

b. $a = 2400/240 = 10$. Therefore,

$$N_p = aN_s = 10 \times 60 = 600 \text{ turns}$$

c. From Eq. 7.6,

$$\phi_m = \frac{2400}{4.44 \times 60 \times 600} = 0.015 \text{ Wb}$$

Although the true transformation ratio is constant (Eq. 7.9), strictly speaking the ratio of the terminal voltages varies somewhat depending on the load and its power factor, because transformers have some losses and are not ideal. In practice, the transformation ratio is obtained from the nameplate data, which lists the primary and secondary voltages under full-load condition, which gives a very good approximation to the turns ratio because the losses are small.

When the secondary voltage V_s is reduced compared to the primary voltage, the transformation is said to be a *step-down transformer*; conversely, if this voltage is raised, it is called a *step-up transformer.* In a step-down transformer the transformation ratio a is greater than unity ($a > 1$), while for a step-up transformer it is smaller than unity ($a < 1.0$). In the event $a = 1$, the secondary voltage equals the primary voltage. This is a special transformer used in instances where electrical isolation is required between the primary and secondary circuit while maintaining the same voltage level. Therefore, this transformer is generally known as an *isolation transformer*.

As is apparent, it is the magnetic flux in the core that forms the connecting link between primary and secondary circuits. In Section 7.5 it is shown how the primary winding current adjusts itself to the secondary load current when the transformer supplies a load.

Looking into the transformer terminals from the source, an impedance is seen which by definition equals V_p/I_p. From Eq. 7.10 we have $V_p = aV_s$ and $I_p = I_s/a$. In terms of V_s and I_s the ratio of V_p to I_p is

$$\frac{V_p}{I_p} = \frac{aV_s}{I_s/a} = \frac{a^2 V_s}{I_s}$$

But V_s/I_s is the load impedance Z_L, thus we can say that

$$Z_{in} \text{ (primary)} = a^2 Z_L \tag{7.11}$$

Secondary (load)

This equation tells us that when an impedance is connected to the secondary side, it appears from the source as an impedance having a magnitude which is a^2 times its actual value. We say that the load impedance is *reflected* or *referred* to the primary. It is this property of transformers that is used in impedance-matching applications. More is said about transferring impedance values in Section 7.6.

7.5   Transformer Under Load

Figure 7.7 shows a transformer supplying a load. The primary and secondary voltages shown have similar polarities, as indicated by the "dot-marking" convention. The dots near the upper ends of the windings have the same meaning as in circuit theory; the marked terminals have the same polarity. Thus when a load is connected to the secondary, the instantaneous load current is in the direction shown. In other words, the polarity markings signify that when positive current enters both windings at the marked terminals, the MMFs of the two windings add.

Since the secondary voltage depends on the core flux ϕ_c, it must be clear that the flux should not change appreciably if E_s is to remain essentially constant under normal loading

marked terminals have same polarity

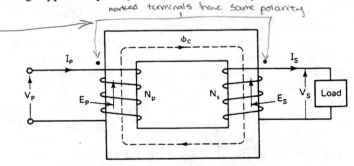

FIGURE 7.7.   Transformer supplying a load.

conditions. With the load connected, a current I_s will flow in the secondary circuit, because the induced EMF E_s will act as a voltage source. The secondary current produces an MMF $N_s I_s$ that creates a flux. This flux has such a direction that at any instant in time it opposes the main flux that created it in the first place. Of course, this is Lenz's law in action. If we assume that ϕ_c increases, the current I_s must have the direction shown as indicated in Figure 7.7 if its resulting flux is to oppose the core flux. Thus the MMF represented by $N_s I_s$ tends to reduce the core flux ϕ_c. This means the flux linking the primary winding reduces and consequently the primary induced voltage E_p. This reduction in induced voltage causes a greater difference between the impressed voltage and the counter induced EMF, thereby allowing more current to flow in the primary. The fact that the primary current I_p increases means that the two conditions stated earlier are fulfilled: (1) the power input increases to match the power output, and (2) the primary MMF increases to offset the tendency of the secondary MMF to reduce the flux.

In general, it will be found that the transformer reacts almost instantaneously to keep the resultant core flux essentially constant. Moreover, the core flux ϕ_c drops very slightly between no load and full load (about 1 to 3%), a necessary condition if E_p is to fall sufficiently to allow an increase in I_p.

The no-load phasor diagram of Figure 7.5b can now be extended so that it includes the load current. This is shown in Figure 7.8, for a secondary current I_s assumed to lag the secondary voltage by the load power factor angle θ. On the primary side, I_p' is the current that flows in the primary to balance the demagnetizing effect of I_s. Its MMF $N_p I_p'$ sets up a flux linking the primary only. Since the core flux ϕ_c remains constant, I_o must be the same current that energizes the transformer at no load. The primary current I_p is therefore the vectorial sum of the currents I_p' and I_o.

Because the no-load current is relatively small, it is correct to assume that the primary ampere-turns equal the secondary ampere-turns, since it is under this condition that the core flux is essentially constant. Thus $N_p I_p = N_s I_s$, which yields Eq. 7.10. Thus we will assume that I_o is negligible, as it is only a small component of the full-load current.

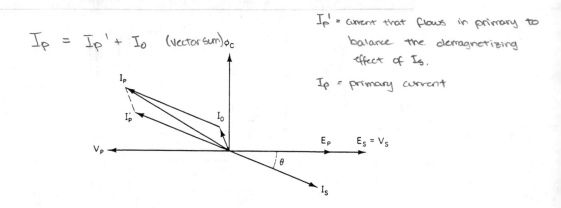

$$I_p = I_p' + I_o \quad \text{(vector sum)}_{\phi_c}$$

$I_p' =$ current that flows in primary to balance the demagnetizing effect of I_s.

$I_p =$ primary current

FIGURE 7.8.   Elementary phasor diagram of a transformer with load; only core losses are considered in this diagram.

Leakage Fluxes

When a current flows in the secondary winding, the resulting MMF $(N_s I_s)$ creates a flux components apart from the flux ϕ_c produced by I_o, which links the secondary winding only. This flux does not link with the primary winding and is therefore not a mutual flux. In addition, the load current that flows through the primary winding creates a flux component which links the primary winding only; it is called the *primary leakage flux*. Figure 7.9 illustrates these fluxes. The *secondary leakage flux* gives rise to an induced voltage that is not counterbalanced by an equivalent induced voltage in the primary. Similarly, the voltage induced in the primary is not counterbalanced in the secondary winding. Consequently, these two induced voltages behave like voltage drops, generally called *leakage reactance voltage drops*. Furthermore, each winding has some resistance which produces a resistive voltage drop. When taken into account, these additional voltage drops would complete the equivalent circuit diagram of a practical transformer, as shown in Figure 7.10. Note that magnetizing branch is shown in this circuit, which for our purposes will be disregarded. This follows our earlier assumption that the no-load current is assumed negligible in our calculations. This is further justified in that it is rarely necessary to predict transformer performance to such accuracies. Since the voltage drops are all directly proportional to the load current, it means that at no-load conditions there will be no voltage drop in either winding.

7.6   Approximate Equivalent Circuit

Although the equivalent circuit for the transformer as shown in Figure 7.10 produces accurate results, it is readily appreciated that for routine calculations it is cumbersome to use. Fortunately, the circuit can be simplified with no loss in accuracy from a practical point of view.

When performing calculations on transformers it is more convenient to combine the resistance and reactance voltage drops that occur on the primary and secondary sides into

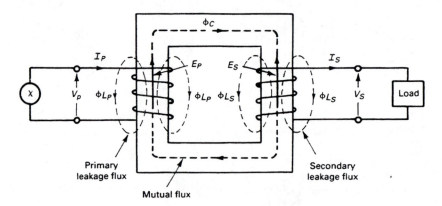

FIGURE 7.9.   Leakage fluxes.

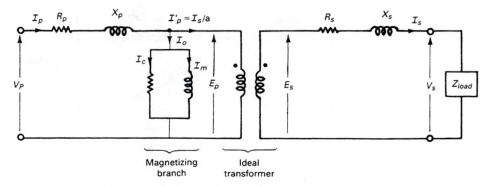

FIGURE 7.10. **Transformer equivalent circuit diagram.**

a single IR and IX value. Obviously, when one side is the high-voltage side and the other the low-voltage side, their respective voltage drops cannot be added numerically.

Figure 7.11a shows the first step to simplify the equivalent circuit diagram of Figure 7.10. We have seen that secondary impedances may be reflected across the transformer to the primary side. Since in doing this the transformation ratio is accounted for, the ideal transformer is also deleted since its inclusion does not serve much purpose. In the process we have deleted the magnetizing branch (which actually accounts for the the magnetizing current), based on our earlier stated assumptions. In this circuit the load voltage reflected to the primary side is aV_s across the reflected load impedance a^2Z_L. If, in addition, the resistive and reactive elements are combined, Figure 7.11b results. This is the simplified equivalent circuit for the transformer in primary terms, in which

$$R_{ep} = R_p + a^2 R_s \quad \Omega$$

$$\text{in primary terms} \tag{7.12}$$

$$X_{ep} = X_p + a^2 X_s \quad \Omega$$

The equivalent circuit, where all quantities are referred to the secondary side of the transformer, is shown in Figure 7.11c, where the equivalent resistance and reactance values in secondary terms are

$$R_{es} = R_s + \frac{R_p}{a^2} \quad \Omega$$

$$\text{in secondary terms} \tag{7.13}$$

$$X_{es} = X_s + \frac{X_p}{a^2} \quad \Omega$$

The voltage V_p/a is the primary terminal voltage reflected to the secondary side. These equivalent circuits are generally more than adequate for solving the relationships between terminal voltages. Finally, note the corresponding phasor diagrams for the simplified circuits.

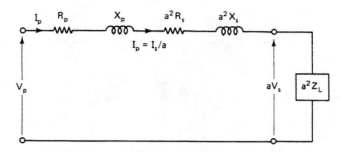

$$I_p = I_s/a$$

(a) Secondary parameters transferred to primary

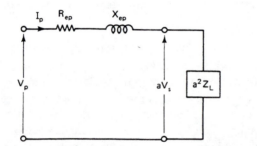

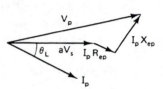

(b) In primary terms

(c) In secondary terms

FIGURE 7.11.   Development of simplified transformer circuit: (a) secondary parameters transferred to primary; (b) in primary terms; (c) in secondary terms.

EXAMPLE 7.2

A 20-kVA 2400/240-V 60-Hz transformer has the following resistance and reactance values:
$R_p = 0.8\ \Omega$ $X_p = 3.0\ \Omega$ $R_s = 0.0084\ \Omega$ $X_s = 0.028\ \Omega$
Calculate the equivalent transformer values

a. in primary terms,

b. in secondary terms.

Solution

a. In primary terms the transformation ratio

$$a = \frac{2400}{240} = 10$$

From Eq. 7.12 we have

$$R_{ep} = R_p + a^2 R_s = 0.8 + 100 \times 0.0084 = 1.64 \ \Omega$$

and

$$X_{ep} = X_p + a^2 X_s = 3.0 + 100 \times 0.028 = 5.8 \ \Omega$$

b. In secondary terms, from Eq. 7.13,

$$R_{es} = 0.0084 + \frac{0.8}{100} = 0.0164 \ \Omega$$

and

$$X_{es} = 0.028 + \frac{3.0}{100} = 0.058 \ \Omega$$

EXAMPLE 7.3

For the transformer in Example 7.2, calculate the equivalent resistive and reactive voltage drops

a. in primary terms, and

b. in secondary terms.

Solution

a. The primary current

$$I_p = \frac{20,000}{2400} = 8.33 \ \text{A}$$

Therefore,

$$I_p R_{ep} = 8.33 \times 1.64 = 13.66 \ \text{V}$$

and

$$I_p X_{ep} = 8.33 \times 5.8 = 48.31 \ \text{V}$$

b. The secondary current

$$I_s = a I_p = 10 \times 8.33 = 83.3 \ \text{A}$$

Therefore,

$$I_s R_{es} = 83.3 \times 0.0164 = 1.366 \ \text{V}$$

and

$$I_s X_{es} = 83.3 \times 0.058 = 4.831 \ \text{V}$$

Note the transformation ratio $a = 2400/240 = 10$, which indicates that the voltage drops can be readily transformed to either side of the transformer.

7.7   *Determination of Equivalent Parameters*

The equivalent parameters in the transformer circuit diagrams representing the transformer may in practice be determined from design data, or failing that, from test data. The latter is an easier way and requires two tests: (1) the open-circuit test, and (2) the short-circuit test.

Open-Circuit Test

In this test, rated voltage is applied to one winding, usually the low-voltage winding for safety reasons, while the other is left open-circuited. The input power supplied to the transformer represents mainly core losses. Since the no-load current is relatively small, the copper losses may be neglected during this test. Therefore, the power input measured is taken as the core loss and remains practically constant at all loads. The circuit instrumentation is shown in Figure 7.12. The W-meter indicates the core loss. The V-meter will register the rated voltage, which in conjunction with the A-meter reading will provide the necessary data to obtain the informaton about the magnetizing branch, if desired.

The core loss can be measured on either side of the transformer. For instance, if a 2400/240-V transformer were to be tested, the voltage would be applied to the secondary side, since 240 V is more readily available. The core loss measured on either side of the transformer would be the same, because 240 V is applied to a winding that has fewer turns than does the high-voltage winding. Thus the volt/turn ratio is the same. This implies that the value of the maximum flux in the core is the same in either case, upon which the core loss depends.

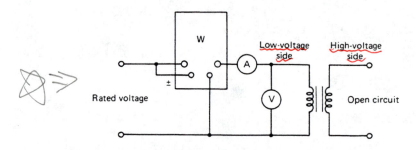

FIGURE 7.12.   Open-circuit test connections.

Short-Circuit Test

The second test required to determine the transformer equivalent circuit parameters is the short-circuit test, for which the connection diagram is shown in Figure 7.13a. As illustrated, the low-voltage side of the transformer is short-circuited. The voltage on the high-voltage winding is adjusted such that rated current flows through the A-meter. In this condition the impedance of the transformer is merely the equivalent impedance, as shown in Figure 7.13b. Again, performing the test on the high-voltage winding is one of convenience since the applied voltage is but a few percent of rated value. To refer to our example of the 2400/240-V transformer, it is easier and more accurate to deal with 5% of 2400 V = 120 V than with 5% of 240 V = 12 V.

With the primary voltage greatly reduced, the flux will be reduced to the same extent. Since the core loss is somewhat proportional to the square of the flux, it is practically zero. Thus a W-meter used to measure the input power will register the copper losses only; the output power is zero. From the input data of watts, amperes, and volts, the equivalent reactance can be calculated, all in terms of the high-voltage side:

$$R_{eH} = \frac{P_{sc}}{I_{sc}^2} \quad \Omega \qquad (7.14)$$

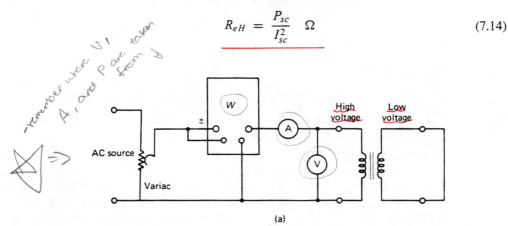

(a)

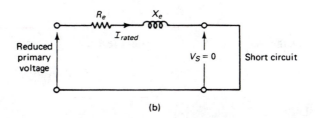

(b)

FIGURE 7.13. **Short-circuit test for a transformer: (a) connection diagram; (b) equivalent circuit.**

$$Z_{eH} = \frac{V_{sc}}{I_{sc}} \ \Omega \qquad (7.15)$$

and

$$X_{eH} = \sqrt{Z_{eH}^2 - R_{eH}^2} \ \Omega \qquad (7.16)$$

where I_{sc}, V_{sc}, and P_{sc} are the short-circuit amperes, volts, and watts, respectively.

EXAMPLE 7.4

The following data were obtained in a short-circuit test on a 20-kVA 2400/240-V 60-Hz transformer:

$$V_{sc} = 72 \text{ V} \quad I_{sc} = 8.33 \text{ A} \quad P_{sc} = 268 \text{ W}$$

Instruments were placed in the high-voltage side with the low-voltage side short-circuited. Obtain the equivalent transformer parameters referred to the high-voltage side.

Solution

$$R_{eH} = \frac{P_{sc}}{I_{sc}^2} = \frac{268}{8.33^2} = 3.86 \ \Omega$$

$$Z_{eH} = \frac{V_{sc}}{I_{sc}} = \frac{72}{8.33} = 8.64 \ \Omega$$

and

$$X_{eH} = \sqrt{Z_{eH}^2 - R_{eH}^2} = \sqrt{8.64^2 - 3.86^2} = 7.73 \ \Omega$$

All measurements in Example 7.4 were taken on the high-voltage side, resulting in equivalent values referred to the high-voltage side of the transformer. However, if the transformer is used as a step-up transformer, it may be necessary to express the equivalent values in terms of the low-voltage side. This is readily obtained without having to perform additional tests. Referring to the example, the transformation ratio $a = 10$ (2400/240). Realizing that, for example, $R_{eH} = R_p + a^2 R_s$, it is readily verified that with $a = 10$, the equivalent value transformed to the low-voltage side in this example is

$$R_{eL} = \frac{R_{eH}}{a^2} = \frac{R_p + a^2 R_s}{a^2} = R_s + \frac{R_p}{a^2}$$

Thus the low-voltage side equivalent resistance becomes

$$R_{eL} = \frac{R_{eH}}{a^2} = 0.0386 \ \Omega$$

Similarly,

$$X_{eL} = \frac{X_{eH}}{a^2} = 0.0773 \ \Omega$$

and

$$Z_{eL} = \frac{Z_{eH}}{a^2} = 0.0864 \ \Omega$$

Therefore, only one short-circuit test needs to be performed, yet by a simple calculation, the equivalent circuit is referred to either side.

7.8   *Voltage Regulation*

The *voltage regulation* of a transformer is, by definition, the difference between the no-load and full-load secondary voltage expressed as a percentage of the full-load voltage. As we may recall, this is similar to the definition for voltage regulation of dc and ac generators. There is an additional condition in the use of transformers that must be fulfilled—that the primary or applied voltage must remain constant. Also, the power factor of the load must be stated since, as can be seen from the phasor diagrams developed for the transformer, the voltage regulation does depend on the load power factor, as with synchronous generators. In general,

$$\text{voltage regulation} = \frac{V_{\text{no load}} - V_{\text{load}}}{V_{\text{load}}} \times 100\% \qquad (7.17)$$

The numerical values employed in the calculations depend on which winding is used as a reference for the equivalent circuit. Similar results are, of course, obtained whether all impedance values are transferred to the primary or secondary side of the transformer.

EXAMPLE 7.5

Using the transformer data of Example 7.4, calculate the voltage regulation of the transformer for an 0.80 lagging PF load.

Solution   The regulation will be calculated in term of primary values. At rated output, the secondary full-load voltage referred to the primary is $I_{sc} = I_p$

$$aV_s = 2400 \text{ V} \qquad I_p = 8.33 \text{ A}$$

The transformer equivalent values determined in Example 7.4, are

$$R_{eH} = 3.86 \ \Omega, \quad X_{eH} = 7.73 \ \Omega \text{ and } Z_{eH} = 8.64 \ \Omega$$

Then

$$\mathbf{Z}_{eH} = Z_{eH}\angle\beta = 8.64\angle\tan^{-1}(7.73/3.86) = 8.64\angle63.5°$$

For a PF = 0.80 lagging ($\theta = \cos^{-1}0.80 = 37°$), the phasor diagram for this example is given in Figure 7.14, with the voltage taken as the reference phasor. Therefore,

$$\begin{aligned}
V_p &= a \, V_s\angle0 + I_p\angle - \theta \times Z_{eH}\angle\beta \\
&= 2400 + 8.33\angle - 37° \times 8.64\angle63.5° \\
&= 2464 + j32 = 2464\angle26.5°
\end{aligned}$$

and the voltage regulation is

$$\frac{2464 - 2400}{2400} \times 100 = 2.67 \ \%$$

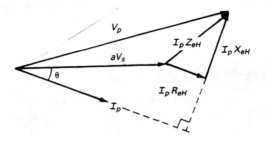

FIGURE 7.14.   Phasor diagram for Example 7.5.

When the calculations are referred in terms of secondary values, we obtain $V_s = 240$ V, $I_p = 83.3$ A and $Z_{e_L} = 0.0864\angle 63.5°\Omega$. Thus, $V'_p = 240\angle 0° + 83.3\angle - 37° \times 0.0864\angle 63.5° = 246.4\angle 26.5°$ V and the voltage regulation is,

$$\frac{246.4 - 240}{240} \times 100 = 2.67\%$$

Hence, identical results are obtained on either side of the transformer. Note that $V_p = aV'_p = 2464$ V, as before.

7.9   Efficiency

The efficiency of a transformer is high and typically in the range 95 to 98%. This implies that the transformer losses are as low as 2 to 5% of the input power. In calculating the efficiency, it is generally much better to determine the transformer losses rather than measuring the input and output powers directly. This is readily appreciated when dealing with large transformers, since the power handled is far too great to apply merely for the purpose of a test. However, the efficiency may be calculated with the use of the equivalent circuit for any given output. As we have seen, the open-circuit test yields the core losses, while the short-circuit test provides the copper losses. Thus the efficiency can be determined from these data with reasonable accuracy. As discussed previously, the core loss consists of two component losses, hysteresis and eddy current loss. The hysteresis loss is due to the cyclic variation of the magnetic flux in the ferrous material. The eddy current loss occurs because of the changing flux in the core, which according to Lenz's law induces a voltage in the core. As a result, circulating currents will be set up with a subsequent I^2R loss. It is thus evident that both of these loss components depend on the frequency and flux. The flux is relatively constant as we discussed earlier. It is therefore assumed that the core losses remain essentially constant throughout the operating range of the transformer. Naturally the frequency is assumed constant, which is the case in practice. In general, the efficiency of any electrical apparatus is

$$\eta = \frac{\text{output power}}{\text{input power}} = \frac{\text{output power}}{\text{output power} + \text{losses}} \tag{7.18}$$

where η is the symbol used to denote efficiency. When Eq. 7.18 is multiplied by the factor 100, the efficiency will then be in percent. Transformer ratings are based on output kVA (MVA), therefore, this equation may also be written as,

$$\eta = \frac{\text{kVA}_{out} \times \text{PF}}{\text{kVA}_{out} \times \text{PF} + \text{copper losses} + \text{core loss}} \tag{7.19}$$

where the copper losses and core loss is then expressed in kW.

The core loss in a transformer is a constant loss for all load conditions. The copper losses vary proportionally to the square of the current. It is therefore a simple matter to calculate the efficiency versus load current curve. Because of the relationship above, it can be shown that the efficiency of a transformer is at a maximum when the fixed losses are equal to the variable losses. In other words, when the copper losses equal the iron losses, the efficiency is maximum.

EXAMPLE 7.6

A 10-kVA 2400/240-V 60-Hz transformer was tested with the following results: power input during short-circuit test = 340 W, power input during open-circuit test = 168 W. Determine

a. the efficiency of this transformer at full load,

b. the load at which maximum efficiency occurs.

The load power factor is 0.8.

Solution

a.

$$\eta_{FL} = \frac{P_{out}}{P_{in}} = \frac{10,000 \times 0.8}{10,000 \times 0.8 + 340 + 168} \times 100 = 94.0\%$$

b. Maximum efficiency occurs when the copper loss = core loss. Thus the copper loss = 168 W. This occurs at a load smaller than full load, in fact, the load current will now be $I^2 R_{eq}/I'^2 R_{eq} = 340/168$ or $I' = I\sqrt{168/340}$. This is the factor by which the power decreases, since $P \propto I^2$, hence

$$P = \sqrt{\frac{168}{340}} \times (10,000 \times 0.8) = 5621 \text{ W} \quad \text{or} \quad 70.3\% \text{ of full load}$$

Therefore,

$$\eta_{max} = \frac{5621}{5621 + 168 + 168} \times 100 = 94.36\%$$

Appendix B gives a computer program to calculate the efficiency of this transformer throughout its entire load range.

7.10   *All-Day Energy Efficiency*

As mentioned in the previous section, transformer loads are generally variable. A transformer connected to the utility that supplies residential power is connected to the power system for 24 hours a day and operates well below the rated power output for most of the time. It is therefore desirable to design a distribution transformer for maximum efficiency occurring at the average output power.

A figure of merit that will more appropriately describe the efficiency performance is the *all-day* or *energy* efficiency η_e of the transformer. We define this as follows,

$$\eta_e = \frac{\text{energy output over 24 hours}}{\text{energy input over 24 hours}} \tag{7.20}$$

Knowing the load cycle of the transformer for a typical 24-hour period enables us to determine the energy efficiency. The unit of kilowatt-hour is still in common use for calculating energy for this purpose.

EXAMPLE 7.7

A 500-kVA 2300/208-V transformer has a core loss $P_c = 3800$ W and a copper loss $P_{cu} = 8200$ W at rated load. The transformer load cycle is:

% load	0	40	75	100	110
Power factor (lagging)	n/a	0.75	0.80	0.86	0.82
Hours	6	5	6	4	3

Determine the all-day energy efficiency of this transformer.

Solution

Energy output in 24 hours $= (0.40 \times 500 \times 0.75 \times 5)$
$$+ (0.75 \times 500 \times 0.80 \times 6) + (1.00 \times 500 \times 0.86 \times 4)$$
$$+ (1.10 \times 500 \times 0.82 \times 3) = 5623 \text{ kWh}$$

Energy losses over 24 hours:

Core loss $= 3.8 \times 24 = 91.2 \text{ kWh}$
Copper loss $= (0.40^2 \times 8.2 \times 5) + (0.75^2 \times 8.2 \times 6)$
$$+ (1.0^2 \times 8.2 \times 4) + (1.1^2 \times 8.2 \times 3) = 96.80 \text{ kWh}$$

Total energy loss $= 91.2 + 96.8 = 188 \text{ kWh}$

$$\eta_e = \frac{5623}{5623 + 188} \times 100 = 96.76 \%$$

7.11 Autotransformers

The principles discussed for the transformer are substantially the same for an autotransformer. It does differ from it in the way the primary and secondary windings are interrelated. Recall that in the discussion of transformer operation a counter EMF was induced in the winding which acted as primary. The induced voltage per turn was the same in each and every turn linking with the common flux in the core. Therefore, fundamentally it makes no difference in the operation whether the secondary induced voltage is obtained from a separate winding linked with the core, or from a portion of the primary turns. The same voltage transformation results in the two situations. When the primary and secondary voltages are derived from the same winding, the transformer is called an *autotransformer*.

An ordinary two-winding transformer may also be used as an autotransformer by connecting the two windings in series and applying the impressed voltage across the two, or merely to one of the windings. It depends on whether it is desired to step the voltage down or up, respectively. Figure 7.15 shows these connections. Considering Figure 7.15a, the input voltage V_1 is connected to the complete winding $(a - c)$ and the load R_L is across a portion of the winding, that is $(b - c)$. The voltage V_2 is related to V_1 as in a conventional two-winding transformer, namely,

$$V_2 = V_1 \times \frac{N_{bc}}{N_{ac}} \tag{7.21}$$

where N_{bc} and N_{ac} are the number of turns on the respective windings. The ratio of voltage transformation in an autotransformer is the same as that for an ordinary transformer, thus

$$a = \frac{N_{ac}}{N_{bc}} = \frac{V_1}{V_2} = \frac{I_2}{I_1} \tag{7.22}$$

with $a > 1$ for step-down. Assuming a resistive load for convenience, $I_2 = V_2/R_L$. If, further, the assumption is made that the transformer is 100% efficient, the power output is

$$P_L = V_2 I_2 \tag{7.23}$$

Note that I_1 flows in the portion of winding ab, whereas the current $(I_2 - I_1)$ flows in the remaining portion bc. The resulting current flowing in the winding bc is always the arithmetic difference between I_1 and I_2, since they are always opposite. Remember that the induced voltage in the primary opposes the primary voltage. As a result, the current caused by the induced voltage flows opposite to the input current. In an autotransformer, the secondary current is this induced current, that is,

$$I_1 + (I_2 - I_1) = I_2 \tag{7.24}$$

Hence the ampere-turns due to section bc, where the substitutions $I_2 = aI_1$ and $N_{bc} = N_{ac}/\alpha$ are made according to Eq. 7.22, is

$$\text{ampere-turns due to } bc = (I_2 - I_1)N_{bc}$$
$$= (aI_1 - I_1)\frac{N_{ac}}{a} = I_1 N_{ac}(1 - \frac{1}{a}) = I_1 N_{ab}$$
$$= \text{ampere-turns due to } ac$$

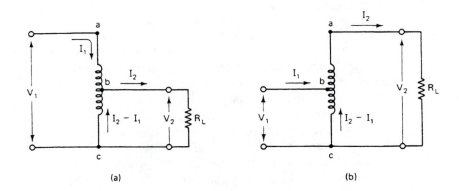

FIGURE 7.15.   Autotransformers: (a) step-down; (b) step-up.

Thus the ampere-turns due to section bc and ac balance each other, a characteristic of all transformer actions.

Equation 7.23 gives the power determined by the load. To see how this power is delivered, we can write this equation in a slightly modified form. By substituting Eq. 7.24 into Eq. 7.23, the following result is obtained,

$$P_L = V_2[I_1 + (I_2 - I_1)]$$
$$= V_2 I_1 + V_2(I_2 - I_1) \quad \text{W} \tag{7.25}$$

This indicates that the load power consists of two parts. The first part is

$$P_c = V_2 I_1 = \text{ conducted power to load through } ab \tag{7.26}$$

The second part is

$$P_{tr} = V_2(I_2 - I_1) = \text{ transformed power to load through } bc \tag{7.27}$$

We will see in the following examples that most of the power to the load is directly conducted by winding ab. The remaining power is transformed by the common winding bc. To show these powers P_c and P_t in terms of the total power P_L, we proceed as follows,

$$\frac{P_c}{P_L} = \frac{V_2 I_1}{V_2 I_2} = \frac{I_1}{I_2} = \frac{1}{a} \tag{7.28}$$

and

$$\frac{P_{tr}}{P_L} = \frac{V_2(I_2 - I_1)}{V_2 I_2} = \frac{I_2 - I_1}{I_2} = \frac{a - 1}{a} \tag{7.29}$$

Thus

$$P_c = \frac{P_L}{a} \quad \text{and} \quad P_{tr} = \frac{P_L(a - 1)}{a} \tag{7.30}$$

with $a > 1$ for a step-down autotransformer.

EXAMPLE 7.8

A standard 5-kVA 2300/230-V distribution transformer is connected as an autotransformer to step down the voltage from 2530 V to 2300 V. The transformer connection is as shown in Figure 7.15a, the 230-V winding is section ab, the 2300-V winding is section bc. Compare the kVA rating of the autotransformer with that of the original two-winding transformer. Also, calculate P_c, P_{tr}, and the currents.

Solution   The rated current in the 230-V winding (or in ab) is

$$I_1 = \frac{5000 \text{ VA}}{230} = 21.74 \text{ A}$$

The rated current in the 2300-V winding (or in bc) is

$$I_2 - I_1 = \frac{5000}{2300} = 2.174 \text{ A} \quad I_2 = 2.174 + I_1 = 23.914 \text{ A}$$

The secondary current I_2 can also be calculated from

$$I_2 = aI_1 = \frac{2530}{2300} \times 21.74 = 23.914 \text{ A}$$

Since the transformation ratio $a = 2530/2300 = 1.1$,

$$P_L = V_1 I_1 = V_2 I_2 = 2530 \times 21.74 = 55.0 \text{ kVA}$$

The conducted power

$$P_c = \frac{P_L}{a} = \frac{55,000}{1.1} = 50 \text{ kVA}$$

and that transformed is

$$P_{tr} = P_L \frac{a-1}{a} = 55,000 \left(\frac{1.1 - 1}{1.1} \right) = 5.0 \text{ kVA}$$

Consider now the step-up autotransformer of Figure 7.15b. Following similar reasons as above, it follows that

$$\begin{aligned} P_L &= V_1 I_1 = V_1 [I_2 + I_1 - I_2)] \\ &= V_1 I_2 + V_1 (I_1 - I_2) \quad \text{W} \end{aligned} \tag{7.31}$$

where we made the substitution of I_1 from Eq. 7.24, which really is Kirchhoff's current law applied to point b. To show this, note at point b we have $I_1 + (I_2 - I_1) = I_2$, so that

$$I_1 = I_2 - (I_2 - I_1) = I_2 + (I_1 - I_2)$$

Again, Eq. 7.31 shows us that the power supplied to the load consists of two parts,

$$P_c = V_1 I_2 = \text{conducted power to load through } ab \tag{7.32}$$

and

$$P_{tr} = V_1(I_1 - I_2) = \text{transformed power to load through } bc \qquad (7.33)$$

In terms of total power we have,

$$\frac{P_c}{P_L} = \frac{V_1 I_2}{V_1 I_1} = \frac{I_2}{I_1} = a \qquad (7.34)$$

and

$$\frac{P_{tr}}{P_L} = \frac{V_1(I_1 - I_2)}{V_1 I_1} = \frac{I_1 - I_2}{I_1} = 1 - a \qquad (7.35)$$

Thus for a step-up autotransformer with $a < 1$, we obtain $P_c = a P_L$ and $P_{tr} = P_L(1 - a)$. As before, P_c is the power directly conducted to the load while P_{tr} is the portion that is transformed.

EXAMPLE 7.9

Repeat Example 7.8 for a 2300- to 2530-V step-up connection, as shown in Figure 7.15b.

Solution   As calculated in Example 7.7, the current rating of the winding ab is $I_2 = 21.74$ A, which also is the load current. The output voltage is 2530 V, thus the VA rating of the autotransformer is

$$P_L = V_2 I_2 = 2530 \times 21.74 = 55 \text{ kVA}$$

which is the same as in Example 7.7. The transformation ratio $a = 2300/2530 = 0.909$, the conducted power is therefore

$$P_c = a P_L = 0.909 \times 55 \text{ kVA} = 50 \text{ kVA}$$

and the transformed power

$$P_{tr} = P_L(1 - a) = 55 \text{ kVA}(1 - 0.909) = 5 \text{ kVA}$$

The examples given make it clear that an autotransformer of given physical dimensions can handle much more load power than an equivalent two-winding transformer. In fact, $a/(a - 1)$ times its rating as a two-winding transformer for the step-down autotransformer or $1/(1 - a)$ for the step-up arrangement. A 5-kVA transformer is capable of taking care of 11 times its rating. These great gains are possible because an autotransformer transforms by transformer action only a fraction of the total power, the power that is not transformed is conducted directly to the load. It should be noted that an autotransformer is not suitable for large percentage voltage reductions as is a power transformer. This is due to the required turns ratio becoming too large, hence the power handling advantage would be minimal. Furthermore, in the unlikely but possible event that the connections to the low-voltage secondary were to fail somewhere below point b in Figure 7.15a, the winding bc would be deleted from the circuit. This implies that the load would see the full high line voltage. For these reasons, autotransformers are not used where large voltage changes are encountered.

In situations were autotransformers can be used to their full advantage, it will be found that they are cheaper than a conventional two-winding transformer of similar rating. They also have better voltage regulation and operate at higher efficiencies. In all applications using autotransformers it should be realized that the primary and secondary circuits are not electrically isolated, since one input terminal is common with one output terminal.

7.12   *Instrument Transformers*

It is often necessary to measure high voltages or large currents. For this it would be desirable if ordinary standard low-range instruments could be used in conjunction with specially constructed accurate-ratio transformers. These are called *instrument transformers* and there are two types: current and potential transformers.

   Current transformers are connected in series in the circuit in which the current is to be measured to step high current values down to a rated value of 5 A for ammeters and wattmeters. A transformer rated 100 to 5 A is designed to take 100 A in its primary winding, and when it does, an A-meter connected to its secondary will indicate 5 A. The current transformer for laboratory use will have multiple connections for various current ratings. For the largest current rating the conductor in which the current is to be measured is fed through the center of the current transformer core to provide a one-turn primary. An additional advantage in the use of these transformers is the isolation from the line, which may be at a high potential.

   To obtain accuracy with a reasonable-sized core, current transformers should never be operated with the secondary open circuited. Unlike power transformers, the number of primary ampere-turns is a fixed quantity for any given primary current. Upon open-circuiting the secondary circuit, the primary MMF is not balanced by a corresponding secondary MMF. This means that all of the primary current becomes excitation current. As a result, a very high flux density is produced in the core, giving rise to a very high secondary induced voltage. The insulation will be subjected to undue high stresses and possible danger to operations. Furthermore, the high magnetizing forces acting on the core may, if suddenly removed, leave behind considerable residual magnetism in the core, so that the ratio obtained after such an open circuit may be appreciably different from that before.

   It is for this reason that even when not in use for measuring purposes, the secondary circuit is closed if the primary current is flowing. There is, of course, no danger when the secondary winding is short circuited, since an ammeter already presents a short circuit. As Figure 7.16 shows, a shorting switch is thus necessary for the removal of the ammeter. Such switches are an integral part of these transformers.

   Potential transformers are specially designed, extremely accurate ratio step-down transformers. They step down the voltage to 120 V so that standard voltmeters and wattmeters, in addition to relays and control devices, may be used. They differ little from ordinary transformers, except that they handle very small amounts of power. A potential transformer used to measure a voltage of 12,000 V will have a transformation ratio of 100:1. The load consists of a high-resistance V-meter, thus the currents are small. For safety rea-

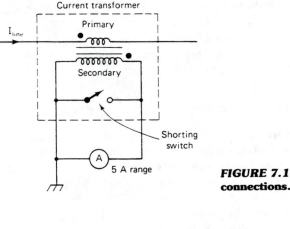

FIGURE 7.16.   **Current transformer connections.**

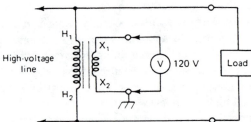

FIGURE 7.17.   **Potential transformer connections.**

sons the secondary side is always grounded and well insulated from the high-voltage side, as shown in Figure 7.17.

7.13   *Three-Phase Transformers*

All electric power generated is distributed by means of a three-phase system. Three-phase power may be transformed by using suitably arranged three individual single-phase transformers or by using a three-phase transformer.

When three separate single-phase transformers are used, we speak of a *transformer bank*. The transformers can be interconnected in four basic configurations: Y-Y, Y-Δ, Δ-Δ, and Δ-Y. These connections follow by considering that the primary windings of the three single-phase transformers may be connected either in wye (Y) or delta (Δ), but so can the secondary windings. With increasing power levels in the transformation of three-phase power, it becomes more and more economical to use a three-phase transformer rather than an arrangement of three separate single-phase transformers.

The unique assembly of the windings on the core of a three-phase transformer makes it possible to save a large amount of iron. By interlinking the magnetic structure the same iron is used by the three phases simultaneously. One other principle advantage is that the costly high-voltage terminals to be brought out the transformer housing are reduced to three, rather than the six necessary for three single-phase transformers. Figure 7.18 shows three core-type transformers placed together to form a single unit having a common path for the return flux. To simplify the illustration, only the primary windings are shown on the outside legs.

If the three transformers are identical, their fluxes will be sinusoidal and balanced. But although they have the same magnitudes, they differ in phase by 120°. Therefore, if the fluxes merge in the common leg, the total resulting flux is,

$$\phi = \phi_1 + \phi_2 + \phi_3$$
$$= \phi_1 \angle 0° + \phi_2 \angle -120° + \phi_3 \angle -240°$$

which will be zero; this implies it is not necessary and can be disregarded. Hence the common leg is omitted and the core-type three-phase transformer manufactured will look like that shown in Figure 7.19. The result is a substantial saving in core material.

Thus for the same kVA rating a three-phase transformer usually costs less than three individual transformers, weighs less, has a higher efficiency, and occupies less space. In addition, fewer connections have to be made to the external circuit.

There are, however, some disadvantages. The three-phase transformer is heavier and bulkier than the individual single-phase transformers, but above all, if one of the phase windings happens to break down in a three-phase transformer, the entire transformer must be removed.

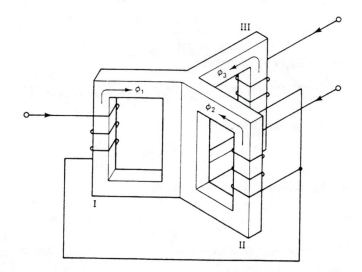

FIGURE 7.18. Core-type three-phase transformer. Only primary windings are shown for clarity.

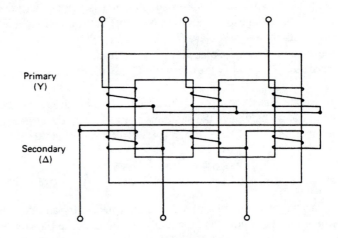

Primary
(Y)

Secondary
(Δ)

FIGURE 7.19.   Three-phase transformer.

EXAMPLE 7.10

A 2400-V three-phase bus supplies a three-phase transformer which delivers 600 kVA to a balanced three-phase 240-V resistive load. Assuming a Y–Y transformation, determine the voltage across each transformer winding and the current through it.

Solution   For a three-phase system, the primary line current is

$$I_{LP} = \frac{\text{kVA} \times 1000}{\sqrt{3} \times V_{LP}} = \frac{600,000}{\sqrt{3} \times 2400} = 144.3 \ \text{A}$$

The secondary line current is

$$I_{Ls} = \frac{\text{kVA} \times 1000}{\sqrt{3} \times 240} = 1443 \ \text{A}$$

The primary winding or phase voltage is

$$V_{ph}(prim) = \frac{2400}{\sqrt{3}} = 1386 \ \text{V}$$

and the secondary winding voltage is

$$V_{ph}(sec) = \frac{240}{\sqrt{3}} = 138.6 \ \text{V}$$

Note that the transformation ratio $a = 2400/240 = 10$. Thus the secondary quantities could also be obtained from

$$I_{Ls} = aI_{Lp} \quad \text{and} \quad V_{ph}(sec) = \frac{V_{ph}(prim)}{a}$$

PROBLEMS

7.1 A single-phase transformer has 400 primary turns and 800 secondary turns. The net iron cross-sectional area of the core is 40 cm². If the primary winding is connected to a 60-Hz supply at 600 V, calculate

    a. the maximum value of the core flux density, and
    b. the secondary induced voltage.

7.2 Determine the rated primary and secondary currents of a 20-kVA 1200/120-V single-phase transformer. If this transformer delivers a load of 12 kW at PF = 0.8, what are the primary and secondary currents?

7.3 A single-phase transformer has 180 and 45 turns, respectively, on its primary and secondary windings. The corresponding resistances are 0.242 Ω and 0.076 Ω, respectively. Determine the equivalent resistance

    a. in primary terms, and
    b. in secondary terms.

7.4 A 60-Hz single-phase transformer has a turns ratio of 8. The resistances are 0.90 Ω and 0.05 Ω and the reactances are 5 Ω and 0.14 Ω for the high- and low-voltage windings, respectively. Determine

    a. the voltage to be applied to the high-voltage side to obtain a full-load current of 180 A in the low-voltage winding, when this winding is short-circuited; and
    b. the PF under the conditions of part (a).

7.5 A 50-kVA 4400/220-V 60-Hz single-phase transformer takes 10.8 A and 544 W at 120 V in a short-circuit test performed on the high-voltage side. Calculate the voltage to be applied to the high-voltage side on full load at a power factor 0.8 lagging when the secondary terminal voltage is 220 V.

7.6 In a 50-kVA 2400/240-V transformer, the iron and copper losses are 680 W and 760 W, respectively. Calculate the efficiency at unity power factor at

    a. full load;
    b. half load.
    c. Determine the load for maximum efficiency.

7.7 A 25-kVA transformer was measured to have a core loss of 480 W at rated voltage. A short-circuit test performed at 125% rated current resulted in a measured power input of 1240 W. Calculate the efficiency of this transformer when it delivers rated kVA at a power factor of 0.78 lagging.

7.8 A short-circuit test was performed on a 10-kVA 2400/240-V transformer with the following data recorded:

$$V_{sc} = 138 \text{ V} \quad P_{sc} = 202 \text{ W} \quad I_{sc} = 4.17 \text{ A}$$

Calculate in primary terms

    a. R_e, Z_e, and X_e, and

    b. the voltage regulation when supplying full load at a power factor of 0.866 lagging.

7.9 A 2300/208-V 500-kVA 60-Hz single-phase transformer was tested by means of the open-circuit test (on the low-voltage winding) and short-circuit test (on the high-voltage winding). The test data obtained are:

    Open-circuit test: $V_{oc} = 208$ V, $I_{oc} = 52.5$ A, $P_{oc} = 3800$ W
    Short-circuit test: $V_{sc} = 95$ V, $I_{sc} = 217.4$ A, $P_{sc} = 6200$ W

Calculate

    a. the efficiency of the transformer at rated kVA and half-rated kVA, when the power factor is 0.866 lagging; and

    b. the voltage regulation when supplying full load at a power factor 0.866 lagging,

    c. maximum efficiency when PF = 0.866.

Use the simplified transformer model of Figure 7.11.

7.10 The magnetic core losses of a 24-kVA 2400/120-V single-phase transformer are 400 W. The copper losses at full load are 900 W. The transformer supplies 80 A at a power factor of 0.80 lagging. Determine the efficiency at this load.

7.11 Determine the all-day efficiency of the transformer in Problem 7.9 on a load cycle supplied at 208 V and approximated by:

           85% full load for 8 hours at a PF = 0.866 lagging
           65% full load for 12 hours at a PF = 0.80 lagging
           15% full load for 4 hours at a PF = 0.75 lagging

7.12 A transformer that can be considered ideal has 200 turns on its primary winding and 500 turns on its secondary winding. The primary is connected to a 220-V sinusoidal supply and the secondary supplies 10 kVA to a load.

    a. Determine the load voltage, secondary current, and primary current.

    b. Find the magnitude of the load impedance as seen from the supply.

7.13 A 20:5 current transformer is connected to a 5-A ammeter that indicates 4.45 A. What is the line current?

7.14 A 25-kVA 440/220-V 60-Hz distribution transformer has a core loss of 740 W. Maximum efficiency occurs when the load is 15 kVA. Determine the efficiency at full load when the power factor equals unity.

7.15 A 75-kVA 6600/230-V transformer requires 310 V across the primary to circulate full-load current on short circuit; the power absorbed is 1.6 kW. Determine the voltage regulation at a PF = 1.0.

7.16 The magnetic core losses of a 24-kVA 2400/120-V 60-Hz single-phase transformer are 400 W. The copper losses at full load are 900 W. The transformer supplies 85 A at a power factor of 0.82 leading. Determine the efficiency at this load.

7.17 A 10-kVA 2200/460-V transformer is connected as an autotransformer to step up the voltage from 2200 V to 2660 V. Determine the kVA rating as an autotransformer.

7.18 A 125-kVA 2400/240-V transformer has its windings connected in series to form a step-down autotransformer. What will be its kVA rating as an autotransformer?

7.19 Three 10-kVA 3980/460-V transformers are connected in Y–Δ to supply a balanced three-phase load of 18 kW at 460 V at a power factor of 0.8 lagging. Determine

    a. the current in each of the primary and secondary windings of the transformers, and

    b. the primary and secondary line currents.

7.20 Three single-phase transformers are connected in a Δ-Y transformer bank arrangement to step down the voltage from 12600 V to 460 V, and to supply a three-phase load of 55 kVA at a power factor of 0.85 lagging. Determine

    a. the transformation ratio of each transformer,
    b. the kVA and kW load of each transformer,
    c. the load currents, and
    d. the current in the transformer windings.

Induction Motors

8.1 Introduction

Most industrial electric machines are three-phase squirrel-cage induction motors. These motors are rugged, rather inexpensive, and require very little maintenance. They range in size from a few watts to about 10,000 Hp. Large induction motors (usually above 5 Hp) are invariably designed for three-phase operation, the reason being that one desires a symmetrical network loading. Small fractional-horsepower motors are usually of single-phase design, referred to in Chapter 9. The speed of an induction motor is nearly constant, dropping only a few percent from its no-load to full-load operation. Induction motors do have certain disadvantages, however, in that:

1. They require an electronic drive to change speed.
2. They run at low lagging power factors when lightly loaded.
3. The starting current is usually five to seven times full-load (rated) current.

In this chapter the induction motor characteristics are examined.

8.2 General Design Features

Of all the electrical machines, the induction motor with a squirrel-cage rotor is the most commonly used ac motor. Typical design features of this motor are illustrated in Figure 8.1, showing the typical stator and squirrel-cage rotor construction of an induction motor.

The stator is composed of laminations of high-grade steel with slotted inner surface to accommodate a three-phase winding. This winding is essentially the same as that for the synchronous generator (see Figure 5.14). The rotor winding design varies depending on the need for torque or speed control. However, the most common is shown in Figure 8.1b, where a squirrel-cage winding is shown. It consists of solid copper or aluminum bars embedded in rotor slots; each bar is short-circuited by end rings, as represented schematically in Figure 8.1c. In the smaller motors, the complete rotor winding, including bars, end rings, and fans, is cast in place. In the larger motors the conductors are formed bars, generally copper, inserted in slots in the rotor laminations. There are no slip rings or carbon brushes, making this construction virtually maintenance free.

The rotor conductors may be nearly parallel to the machine axis, or "skewed." This is to provide a more uniform torque and to reduce noise during operation. Furthermore, it prevents the possibility of rotor and stator teeth from lining up opposite each other and locking in place, normally referred to as *cogging action* due to the reluctance torque.

The *squirrel-cage rotor* is built with a very small air gap and is equipped with close-fitting ball bearings rather than sleeve bearings, which must have clearance. As the rotor bars

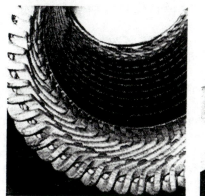

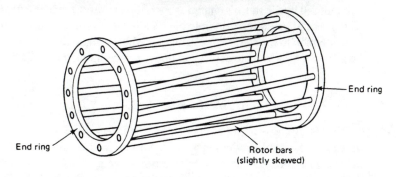

(c)

FIGURE 8.1. Induction motor construction: (a) stator; (b) squirrel-cage rotor; (c) sketch of squirrel-cage winding. [(a),(b) Courtesy of Siemens.]

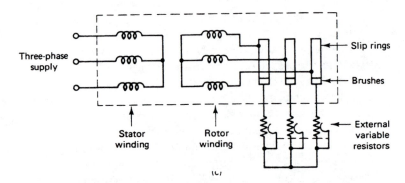

FIGURE 8.2. Schematic diagram of a wound-rotor induction motor.

(conductors) are cut by the stator flux they have a voltage induced in them by transformer action. A current will flow in the short-circuited rotor bars, causing torque to be developed between these currents and the rotating magnetic field of the stator. It should be noted that the squirrel-cage winding adjusts itself to any number of poles, dictated by the number of poles the stator winding is wound for; the number and shape of the rotor bars used is dictated by the desired motor characteristics.

The second type of rotor construction is the *wound rotor*. In this type of construction the rotor has a three-phase winding as well, similar to the stator and wound for the same number of poles as the stator winding. The wound-rotor winding terminates in slip rings mounted on the rotor shaft. Brushes ride on the slip rings, and during the starting period of the motor they are connected externally to a three-phase resistor bank (one in each phase and Y-connected). See Figure 8.2. The external resistances are shorted out simultaneously in one or more steps as the motor comes up to speed. Thus this rotor construction has the added feature of allowing external three-phase resistors to be added in series to the rotor winding for starting purposes and for providing speed control. Details of these operations are discussed in later sections.

Regardless of the rotor construction employed, the rotor currents in an induction motor are induced by the stator's changing, or rather, rotating magnetic field. This induction action is the central operating principle of ac induction motors. Because of this, we will now turn our attention to the stator winding and see how a stationary three-phase winding connected to a three-phase voltage source produces a rotating magnetic field of constant magnitude.

8.3   The Rotating Magnetic Field

When the stator winding is connected to a three-phase voltage supply, currents will flow in each phase of this winding. These currents will be displaced from each other by 120°, as shown in Figure 8.3a. To be consistent, we shall assume that when the instantaneous currents are positive, they agree with the arrows in Figure 8.3b. Furthermore, for positive currents as shown, the currents will enter the phase windings at a, b, and c and be directed into the page. Should a phase current be on the negative half-cycle of its sine wave, the current direction will reverse (i.e., it will be directed out of the page at a, b, and c).

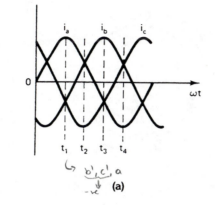

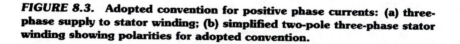

(a)

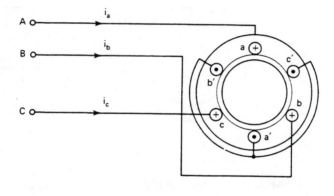

(b)

FIGURE 8.3. Adopted convention for positive phase currents: (a) three-phase supply to stator winding; (b) simplified two-pole three-phase stator winding showing polarities for adopted convention.

Consider the instant of time $t = t_1$ in Figure 8.3a. At this time the current i_a is positive, and i_b and i_c are negative. The resulting current directions in the phase windings at this time t_1 is then as illustrated in Figure 8.4a. The cumulative effect of these currents flowing in the phase windings is that the magnetic field is directed to the left. This is readily confirmed by the right-hand rule. As this procedure is followed for the next three time intervals indicated in Figure 8.3a, the magnetic field will have rotated 180° in the clockwise direction, or one pole span, as Figure 8.4 indicates. This rotation occurred during half a cycle of the supply frequency. Therefore, the field will revolve a distance covered by two poles for each cycle of the supply frequency. This implies that the speed of the rotating magnetic field is inversely proportional to the number of pole pairs and proportional to the frequency. Expressing this in formula form gives what is known as the *synchronous speed* n_s of the rotating magnetic field, namely,

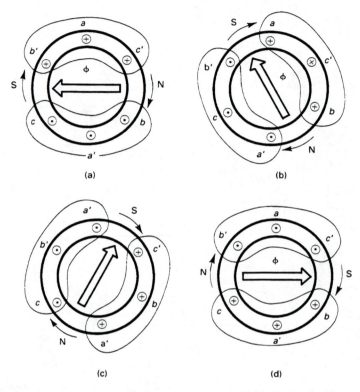

FIGURE 8.4. Rotation of the stator magnetic field during half a cycle of the line frequency. Instants in time indicated refer to those in Figure 8.3a. (a) $t = t_1$; (b) $t = t_2$; (c) $t = t_3$; (d) $t = t_4$.

$$n_s = \frac{f}{P/2} \text{ r/s} = \frac{120f}{P} \text{ r/min} \qquad (8.1)$$

By analytical analysis the concept of the rotating field of constant magnitude can be proven. This analysis is deferred to Appendix C.

It is worth noting that by reversing the phase sequence of the supply, by interchanging any of the two currents in Figure 8.3a, the field will rotate in the counterclockwise direction. This is readily verified. In practice this is the procedure followed to reverse the direction of rotation of an induction motor by interchanging any two line connections at the motor terminals. The ease by which it is possible to reverse the direction of rotation of the motor constitutes one of the advantages of three-phase machines.

8.4 Slip and Rotor Speed

The rotor of an induction motor rotates in the same direction as that of the revolving field. It cannot do so at synchronous speed, otherwise the rotor conductors rotate in unison with the magnetic field, and no flux would be cut. An examination of Figure 8.5 will make this clear. Imagine for a moment that the rotating magnetic field is rotating clockwise and instantaneously directed as shown by ϕ_s. As the field sweeps through a coil (indicated by $a - a'$) on the rotor it will induce an EMF in it. Because of the short-circuited rotor circuit, an induced current is created in this coil which flows in the direction shown, as determined by Lenz's law. We can determine the direction of the developed force (hence, *torque*) on the coil sides a and a'. For example, applying the left-hand rule to coil side a yields a force in the direction represented by F. Thus, indeed, the torque is directed in such a way as to rotate the rotor in the direction of the rotating field, as Figure 8.5 illustrates. The rotor follows the main field in order to catch up to it. As you can appreciate, there must always be relative motion between the rotating field and the rotor conductors, otherwise there will be no induced rotor currents, hence, no torque. The difference between the synchronous speed n_s and the rotor speed n_r is called the *slip s*, and is expressed as a percentage of synchronous speed as:

$$\text{slip} = \frac{n_s - n_r}{n_s} \times 100\% \tag{8.2}$$

The rotor speed may be expressed as

$$n_r = (1 - s) \times n_s \quad \text{r/min} \tag{8.3}$$

where s is expressed as a decimal.

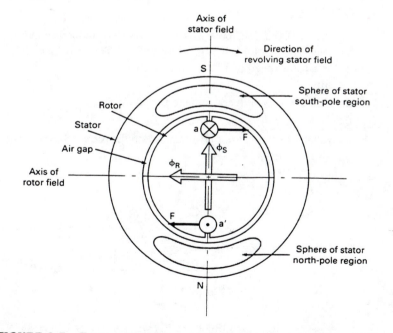

FIGURE 8.5.   Torque development in induction motor schematically predicted for a single rotor coil a–a'.

8.5   *Rotor-Induced Voltage and Frequency*

At standstill, when the rotor is at rest, the rotating field sweeps the rotor bars at its maximum rate. Under those conditions, the generated voltage in the rotor circuit will be maximum and determined by the number of turns on the rotor. As the field revolves, a back EMF is generated in the stator winding which is nearly equal to the impressed voltage. It thus follows that at standstill the flux sweeps the stator turns at the same rate as those in the rotor. This means that the induced voltages in the rotor and stator turns on a per phase basis are related by the turns ratio, as is the case in a transformer between primary and secondary. It also follows that the frequency of the rotor-induced voltage equals the line frequency when the rotor is at rest. In this condition the slip $s = 1.0$ or 100%. As the slip decreases, the rate at which the flux sweeps across the conductors decreases proportionately and the rotor EMF becomes

$$E_R = s \times E_{BR} \qquad\qquad (8.4)$$

and the rotor frequency

$$f_R = s \times f \qquad\qquad (8.5)$$

where E_R = rotor-induced voltage per phase at slip s
$\qquad E_{BR}$ = blocked rotor-induced voltage per phase
$\qquad\quad f_R$ = rotor frequency

EXAMPLE 8.1

A three-phase 60-Hz four-pole 220-V wound-rotor induction motor has a stator winding Δ-connected and a rotor winding Y-connected. The rotor has 40% as many turns as the stator. For a rotor speed of 1710 r/min, calculate
a. the slip,
b. the blocked rotor-induced voltage per phase E_{BR},
c. the rotor-induced voltage per phase E_R,
d. the voltage between rotor terminals when the rotor is blocked and open-circuited, and
e. the rotor frequency.

Solution
a.

$$n_s = \frac{120 f}{P} = \frac{120 \times 60}{4} = 1800 \ \text{r/min}$$

$$s = \frac{n_s - n_r}{n_s} = \frac{1800 - 1710}{1800} = 0.05$$

b.

$$E_{BR} = 40\% \text{ of } V_{stator}/\text{phase} = 0.4 \times 220 = 88 \ \text{V/phase}$$

c. $E_R = sE_{BR} = 0.05 \times 88 = 4.4 \text{ V}$

d. $V_{L-L}(\text{rotor}) = \sqrt{3} \times 88 = 152.4$ V

e. $f_R = sf = 0.05 \times 60 = 3$ Hz

As indicated, under normal running conditions the induced rotor voltage and frequency are quite low.

8.6 The Rotor Circuit

From what has been discussed so far on the three-phase induction rotor, it is apparent that it is essentially a transformer with a short-circuited secondary that is free to move continuously with respect to the primary. It is therefore anticipated that a simplified equivalent circuit diagram resembles that of the single-phase transformer discussed. This is indeed the situation. It has been shown in Section 8.5 that the induced rotor voltage per phase is sE_{BR}. Since this voltage acts in the short-circuited rotor winding, it will set up currents that will be limited only by the rotor impedance. This impedance is made up of two components: (1) the rotor resistance R_R, and (2) the leakage reactance sX_{BR}, where X_{BR} is the rotor reactance at standstill. Since the reactance is a function of frequency, the leakage reactance is proportional to the slip. As a result, the rotor current becomes

$$I_R = \frac{sE_{BR}}{\sqrt{R_R^2 + (sX_{BR}^2)}} \tag{8.6}$$

If both numerator and denominator of Eq. 8.6 are divided by the slip s, we obtain

$$I_R = \frac{E_{BR}}{\sqrt{(R_R/s)^2 + X_{BR}^2}} \tag{8.7}$$

Although this implies a simple algebraic operation, it is of great significance. As Eq. 8.7 indicates, the current I_R can now be considered to be produced by a voltage E_{BR} of line frequency, whereas the current determined by Eq. 8.6 is of slip frequency. In other words, the division by s has changed the point of reference from the rotor circuit to the stator circuit. Translating Eq. 8.7 into an equivalent electric circuit diagram, it becomes as shown in Figure 8.6a.

Since it is convenient to deal with the actual rotor resistance R_R, the term R_R/s is split into two components:

$$\frac{R_R}{s} = \frac{R_R}{s} + R_R - R_R = R_R + R_R\frac{1-s}{s} \tag{8.8}$$

and a corresponding circuit diagram is that of Figure 8.6b. If Eq. 8.8 is now multiplied through by I_R^2, we obtain an equation representing power terms,

$$I_R^2\frac{R_R}{s} = I_R^2 R_R + I_R^2 R_R\frac{1-s}{s} \tag{8.9}$$

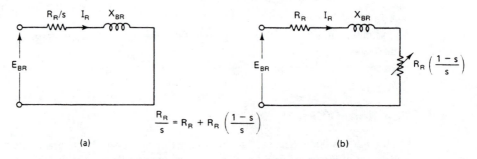

FIGURE 8.6. Rotor circuit diagram on a per phase basis. Resistive element representing rotor copper loss and rotor power developed: (a) combined; (b) separated.

The left-hand side of the equation represents the total power input to the rotor circuit, which is made up of two components; (1) the power dissipated as copper loss in the rotor circuit $I_R^2 R_R$, and (2) the electric power that is converted into mechanical power, $I_R^2 R_R[(1 - s)/s]$. Thus on a per phase basis,

    rotor power input (RPI) = rotor copper loss (RCL) + rotor power developed (RPD)

where

$$RPI = I_R^2 \frac{R_R}{s} \tag{8.10}$$

$$RCL = I_R^2 R_R \tag{8.11}$$

$$RPD = I_R^2 R_R \frac{1 - s}{s} = RPI(1 - s) \tag{8.12}$$

It is interesting to note that the mechanical output power is represented in the electrical circuit by a resistance having a value $R_R(1 - s)/s$. In general, the power developed by a motor is the product of its torque and the angular velocity of the rotor. Therefore,

$$P_d = \omega_R T \quad W$$

It follows, then, that the developed torque by the motor is

$$T_d = \frac{RPD}{\omega_R} \quad N \cdot m \tag{8.13}$$

where $\omega_R = 2\pi n_r/60$ rad/s and n_r is the rotor speed in revolutions per minute at which the power is developed. Later it will be seen that to obtain the output torque, the rotational losses must be accounted for.

8.7 *Complete Circuit Diagram*

So far, the rotor circuit was developed and it was shown that the power transformed across the air gap represents the rotor losses and the mechanical power developed by the motor. To complete the equivalent circuit diagram, the stator circuit must be included. The stator phase winding, having a resistance R_s and a leakage reactance X_s, also has a magnetizing branch. This magnetizing branch, unlike the transformer equivalent circuit, cannot be neglected here, because of the presence of the air gap. However, in this text we will make certain assumptions which will aid significantly in the performance calculations of the motor characteristics without sacrificing accuracy. The stator equivalent circuit can be represented by Figure 8.7a. The following assumptions will now be made to expedite numerical calculations, namely:

1. The core loss is assumed constant and obtained from the no-load test to be described later. This means that the resistance R_c can be deleted from the equivalent circuit diagram. This should not be interpreted as the core loss being negligible, but merely as a constant loss, which in calculating the motor efficiency must be accounted for.

2. As mentioned when discussing the transformer, the magnetizing current was a negligibly small component of rated current. Here it is approximately 30 to 50% of rated current, depending on the size of the motor. Thus the magnetizing reactance is an essential component in the equivalent circuit. However, calculations can be simplified with not much loss in accuracy, by moving the magnetizing reactance to the input terminals as shown in Figure 8.7b.

At this point it remains to combine the rotor and stator circuit diagrams to yield an equivalent diagram, on a per phase basis, for the induction machine. In doing so it must be realized that both circuits must be compatible. That is, the rotor parameters must be referred to the stator side, identical to what was done for the transformer by referring

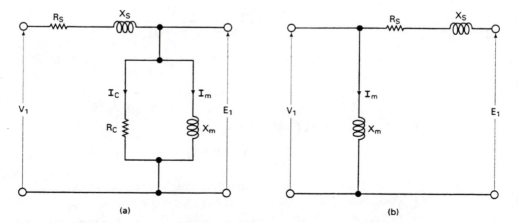

(a)   (b)

FIGURE 8.7. Stator equivalent circuit diagram of induction motor, per phase.

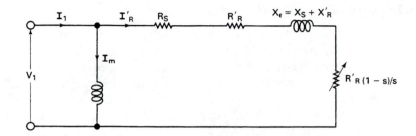

FIGURE 8.8.   Equivalent circuit diagram for the induction motor on a per basis referred to the stator.

secondary quantities to the primary side. In other words, Figure 8.6a or b can be joined to Figure 8.7b, provided that E_{BR} equals E_1. To ensure this equality, the turns ratio between stator and rotor-phase windings must be accounted for, such that $E'_{BR} = a\, E_{BR} = E_1$, where a represents this ratio and E'_{BR} denotes stator-referred rotor quantity. Having done so, the equivalent circuit referred to the stator side is that of Figure 8.8.

From the equivalent diagram it can now be appreciated why induction motors at light load operate under such poor power factors. At light loads (small slip values), the resistor representing the mechanical power $R'_R\,(1-s)/s$ is large. This means that I'_R is relatively small compared to the magnetizing current I_m. Therefore, the circuit behaves largely inductive since X_m is the dominating element in the equivalent circuit and the power factor is therefore small. With increased loading, the resistance $R'_R\,(1-s)/s$ decreases quickly since the slip s increases. Therefore, it rapidly becomes the dominating element in its branch, with a corresponding large improvement in the overall circuit power factor with increasing loads.

The rotor current in Figure 8.8 in stator terms is

$$I'_R \;=\; \frac{\mathbf{V}_1}{(R_s + R'_R/s) + j(X_s + X'_R)} \tag{8.14}$$

which enables calculation of the RPI, RCL, and RPD according to Eqs. 8.11 to 8.12. The magnetizing current I_m is

$$I_m \;=\; \frac{\mathbf{V}_1}{j\,X_m} \tag{8.15}$$

and therefore the motor line current

$$\mathbf{I}_L \;=\; \mathbf{I}_m + \mathbf{I}'_R \tag{8.16}$$

where the boldface type indicates that the currents are phasor quantities. Since the magnetizing branch is moved to the input terminals the stator quantities appear as part of the rotor circuit. This would result in a negligible error if $V_1 \approx E_1$ in Figure 8.7b.

In practical work, this simplification in the normal operating range of the motor leads to insignificant errors and is therefore adopted as explained.

The output or shaft torque is

$$T = \frac{\text{RPD} - \text{mechanical losses}}{\omega_R} \tag{8.17}$$

where the mechanical losses include friction, windage, and core losses, which, as a combined loss, can be obtained from a no-load test yet to be discussed. Normally, these combined losses are known as *rotational loss*.

EXAMPLE 8.2

A three-phase 220-V 60-Hz six-pole 10-Hp induction motor has the following circuit parameters on a per basis referred to the stator:

$$R_S = 0.344 \ \Omega \qquad R'_R = 0.147 \ \Omega$$
$$X_s = 0.498 \ \Omega \qquad X'_R = 0.224 \ \Omega \quad X_m = 12.6 \ \Omega$$

The rotational losses including the core losses amount to 262 W and may be assumed constant. For a slip of 2.8% determine
a. the line current and the power factor,
b. the shaft torque and output horsepower, and
c. the efficiency.

Solution Assuming a Y-connected stator winding, the phase voltage is $220/\sqrt{3} = 127$ V; the equivalent circuit is given in Figure 8.9.
a.

$$I'_R = \frac{127}{0.344 + 5.25 + j0.722} = 22.52\angle -7.4° = 22.33 - j2.88 \ \text{A}$$

$$I_m = \frac{127}{j12.6} = -j10.08 \ \text{A}$$

The line current

$$I_L = 22.33 - j(2.88 + 10.08) = 25.82\angle -30.1° \ \text{A}$$

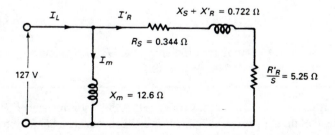

FIGURE 8.9. Circuit diagram for Example 8.2.

and the power factor

$$PF = \cos(-30.1°) = 0.865$$

b.

$$n_s = \frac{120 \times 60}{6} = 1200 \text{ r/min}$$

$$n_r = (1 - 0.028) \times 1200 = 1166 \text{ r/min}$$

$$\omega_R = \frac{2\pi \times 1166}{60} = 122.1 \text{ r/min}$$

$$RPI = \frac{3I_R'^2 R_R'}{s} = 3 \times 22.52^2 \times 5.25 = 7988 \text{ W}$$

$$RPD = RPI(1 - s) = 7988 \times (1 - 0.028) = 7764 \text{ W}$$

$$P_{out} = RPD - P_{rot} = 7764 - 262 = 7502 \text{ W}$$

$$T = \frac{7502}{122.1} = 61.4 \text{ N} \cdot \text{m}$$

$$\text{horsepower} = \frac{7502}{746} = 10.1$$

c. Losses

rotational loss	=	262 W
RCL = 0.028 × 7988	=	224 W
Stator copper loss 3 × 22.52² × 0.344	=	523 W
Total loss	=	1009 W

Therefore, the efficiency

$$\eta = \frac{7502}{7502 + 1009} \times 100 = 88.1\%$$

Note that powers calculated on a per phase basis are multiplied by 3 to obtain the three-phase power value. Also, the input power can be obtained from

$$P_{in} = \sqrt{3}I_L V_L \cos\theta = \sqrt{3} \times 25.82 \times 220 \times 0.865 = 8510.5 \text{ W}$$

So that calculating the efficiency from P_{out}/P_{in} results in the same efficiency, namely, $\eta = 7502/8510.5 = .881$ or 88.1%.

Furthermore, the reader may observe that the stator resistance R_s carries the current I_R' instead of the stator current I_L. This is as a consequence of moving the magnetizing reactance to the input terminals. As mentioned before, this expedites performance calculations which is justified in view of the small errors committed.

With regard to Example 8.2, the calculations may be repeated for any value of slip to obtain the complete performance characteristic as a function of speed. This will be left to the reader as a programming exercise, which is deferred to Problem 8.9 (also, see Appendix D).

8.8   *Induction Motor Characteristics*

For a variable-speed induction motor we may use a wound rotor so that variable external resistances may be used in the rotor circuit to vary the speed, or to increase the starting torque. This type of motor is generally used where frequent starts under load are necessary, such as for hoists and cranes. When fairly constant power is required, starting is infrequent, and only average starting torque is necessary (such as for driving pumps, blowers, and fans), the much simpler and therefore cheaper squirrel-cage motor is normally employed. Figure 8.10 shows a typical characteristic curve for a three-phase induction motor with squirrel-cage rotor.

Common squirrel-cage motors are classified into four specific designs, A through D. Their specific characteristics are defined by the National Electrical Manufacturers Association (NEMA) and result in markedly different machine characteristics. Of the four designs, only B, C, and D are in common use and are listed in manufacturers' handbooks. These four classes, however, cover nearly all practical applications of induction machines.

1. *Class A motors.* These motors are characterized by having a low rotor-circuit resistance and therefore operate at very small slips ($s < 0.01$) under full load. Machines in this class have high starting currents and normal starting torques, because of their low rotor resistance.

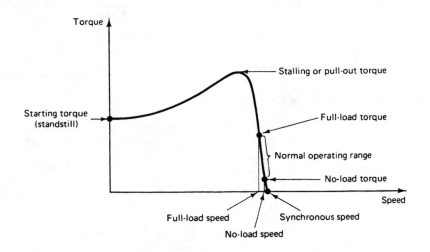

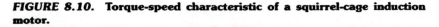

FIGURE 8.10.   Torque-speed characteristic of a squirrel-cage induction motor.

2. *Class B motors.* These are general-purpose motors of normal starting torque and starting current. The speed regulation at full load is low (usually under 5%) and the starting torque is on the order of 150% of rated, being lower for the lower speed and larger motors. It should be realized that although the starting current is normal, it generally is 600% of full-load value.

3. *Class C motors.* Compared to class B motors, class C motors have higher starting torque, normal starting currents, and run at slips of less than 0.05 at full load. The starting torque is about 200% of rated, and the motors are generally designed to start at full load. Typical applications of this class motor are conveyors, reciprocating pumps, and compressors.

4. *Class D motors.* These are high-slip motors with high starting torque and relatively low starting current. As a result of the high full-load slip, their efficiency is generally lower than that of the other motor classes. The peak of the torque characteristic is moved to the zero point of the torque-speed curve, resulting in a starting torque of about 300%, which is identical to the stalling torque.

As discussed, it is evident that the characteristics of an inducton motor can be influenced considerably. How is this accomplished? The answer is simple: by shaping the rotor conductor bars. Figure 8.11 illustrates what is meant; it shows various rotor slot configurations to achieve the desired torque-speed motor characteristics. The rotor construction, depicted in Figure 8.11b, has in effect two squirrel-cage windings. The low-resistance, high-reactance winding is the one that is embedded deeply into the rotor. The top one has a higher resistance (smaller conductor cross section) but a lower reactance (leakage reactance is lower). At startup and lower speeds, the frequency of the current in the rotor bars is high, and the current tends to flow in the upper, low-reactance, high-resistance winding, thereby producing high starting torque. As the motor speed increases and subsequently the rotor frequency decreases, the rotor current will essentially flow in the deeper, low-resistance winding. Figure 8.12 shows the influence of the various parameters on the induction motor characteristics.

8.9 Starting Torque

In Figure 8.10, certain specific points, such as the starting torque and stalling torque, are identified. The starting torque occurs at startup with the rotor at standstill (i.e., the slip $s =$

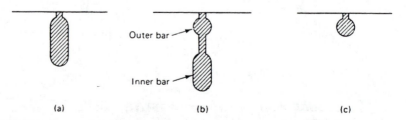

| (a) | (b) | (c) |

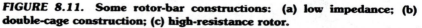

FIGURE 8.11. Some rotor-bar constructions: (a) low impedance; (b) double-cage construction; (c) high-resistance rotor.

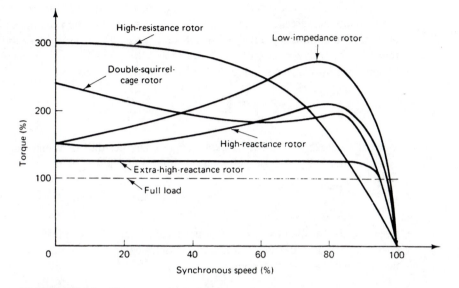

FIGURE 8.12.   Torque-speed characteristic of squirrel-cage induction motors with various rotor constructions.

1.0). If this value of slip is substituted when calculating the rotor current at standstill, we obtain

$$I'_{R(st)} = \frac{V_1}{\sqrt{(R_s + R'_R)^2 + (X_s + X'_R)^2}} \qquad (8.18)$$

Then $\text{RPI}_{st} = 3I'_{R(st)} R'_R$ can be determined, which is the rotor power input at the instant of starting. From this the starting torque can be obtained.

EXAMPLE 8.3

Calculate the starting torque of the motor in Example 8.2.

Solution

$$I'_{R(st)} = \frac{127}{\sqrt{(0.344 + 0.147)^2 + (0.498 + 0.224)^2}} = 145.45 \text{ A}$$

$$\text{RPI}_{st} = 3 \times (145.45)^2 \times 0.147 = 9330 \text{ W}$$

$$T_{st} = \frac{\text{RPI}_{st}}{\omega_s} = \frac{9330}{2\pi \times 1200/60} = 74.2 \text{ N} \cdot \text{m}$$

Maximum Torque

As the load on the motor is increased, the slip increases just enough to meet this new load requirement. This increased loading can generally be carried somewhat beyond the value the motor is designed for, but at the expense of overheating. Assuming that the load on the machine is increased further, there will be a point at which the motor cannot produce any additional torque, and the motor will stall. It can be shown by the familiar impedance-matching principle in circuit theory that this will happen when the impedance of R'_R/s matches the impedance looking back into the source. Accordingly,

$$s_{mt} = \frac{R'_R}{\sqrt{R_s^2 + (X_s + X'_R)^2}} \qquad (8.19)$$

When the stator resistance R_s is small compared to the equivalent reactance $X_e = X_s + X'_R$ (which is normally true for the larger machines), we can say that

$$s_{mt} = \frac{R'_R}{X_e} \qquad (8.20)$$

where s_{mt} is the slip at which maximum torque occurs. Using this value of slip, the rotor current can be found from Eq. 8.14; this enables calculation of the rotor power input at the point of maximum torque, namely,

$$I'_{Rmt} = \frac{V_1}{\sqrt{(R_s + X_e)^2 + X_e^2}} \approx \frac{V_1}{\sqrt{2}\,X_e} \qquad (8.21)$$

and

$$\text{RPI}_{mt} = 3I'^2_{Rmt}\frac{R'_R}{s_{mt}} = \frac{3\,V_1^2}{2\,X_e} \qquad (8.22)$$

which can then be used to determine the rotor-developed power as $\text{RPD}_{mt} = (1 - s)\,\text{RPI}_{mt}$, from which $P_{out} = \text{RPD}_{mt} - P_{rot}$, representing the output at the shaft. The stalling or maximum torque is then obtained from $T_{mt} = P_{out}/\omega_R$. Note that the torque is dependent on V_1^2.

It is interesting to see that the maximum output torque on the shaft of the motor, as derived from Eq. 8.22, is independent of the rotor circuit resistance, R'_R. Increasing the rotor circuit resistance by adding external resistances (as can be done in a wound-rotor induction machine), merely increases the slip (or decreases the speed) at which this maximum torque occurs, in accordance with Eq. 8.19, but its magnitude is unchanged. This is illustrated in Example 8.5.

EXAMPLE 8.4

Calculate the maximum torque that the motor of Example 8.3 can develop and the speed at which this occurs.

Solution   Since R_s is not much smaller than X_e, we must use Eq. 8.19 to calculate the slip at which maximum torque occurs. Therefore,

$$s_{mt} = \frac{0.147}{\sqrt{0.344^2 + (0.498 + 0.224)^2}} = 0.184$$

from which

$$R'_R = \frac{0.147}{0.184} = 0.799 \ \Omega$$

$$n_{r\,mt} = 1200(1 - 0.184) = 979 \ \text{r/min}$$

$$I'_{R\,mt} = \frac{127}{\sqrt{(0.344 + \frac{0.147}{0.184})^2 + (0.498 + 0.224)^2}} = 93.94 \ \text{A}$$

$$RPI_{mt} = 3 \times 93.94^2 \times 0.799 = 21153 \ \text{W}$$

$$RPD_{mt} = RPI_{mt}\,(1 - s_{mt}) = 21153 \times (1 - 0.184) = 17264 \ \text{W}$$

and

$$T_{mt} = \frac{RPD_{mt} - P_{rot}}{\omega_{r\,mt}} = \frac{17261 - 262}{2\pi \times 979/60} = 165.9 \ \text{N·m}$$

Summarizing the results of Examples 8.2, 8.3, and 8.4 we see that the full load torque $T_{FL} = 61.4$ N·m, $T_{st} = 74.2$ N·m, and $T_{mt} = 165.9$ N·m. Therefore, $T_{st}/T_{FL} = 1.21$, and $T_{mt}/T_{FL} = 2.70$, which represent typical ratios for a class B induction motor.

8.10   *Wound-Rotor Motors*

As discussed earlier, the basic difference between the wound-rotor induction motor and the squirrel-cage induction motor is basically in the rotor construction. Both motors use similar stators. The squirrel-cage motor has the limitation of constant rotor resistance. Under normal running conditions a high efficiency is desired, which calls for a low rotor resistance, but this in turn implies high starting currents and low starting torque. Thus the rotor design must be a compromise.

The wound-rotor motor, on the other hand, provides an effective way to circumvent this compromise. This is made possible by connecting the rotor winding to slip rings, which are in contact with brushes. Thus provisions are made to connect external resistances in series with the rotor winding. At startup and with added resistance, the starting current is reduced, the starting torque increased, and the power factor is improved. As is apparent from Eq. 8.19, the slip at which maximum torque occurs is directly related to the rotor resistance. Thus, by adding the appropriate amount of resistance into the rotor circuit the startup torque can be made equal to the maximum torque if a high starting torque is desired. As the motor speeds up, the external resistances are decreased in such a manner as to provide maximum torque throughout the accelerating range. Once the motor reaches operating speed the external resistances are shorted out at the brushes.

Since the maximum torque can be maintained throughout the accelerating range, the wound-rotor motor is desirable when starting high-inertia loads. Furthermore, this motor also has the advantage that during startup the heating effect in the motor is less, since the I^2R losses are dissipated mainly in the external resistances, rather than in the rotor itself. Although this may help the motor, the energy wasted is considerable in large motors.

Therefore, schemes are available, called *slip-recovery systems*, to recover the energy that would otherwise be lost in the external resistances.

To show the effects of varying rotor resistance on the torque-speed characteristics, the following example is included. To simplify the procedure, calculations will be limited to the rotor circuit only, according to Figure 8.6. This is justified, of course, since rotor parameters can be determined by measurement. The transformation ratio can also be obtained, therefore, by referring the rotor quantities in stator terms, including the stator resistance and magnetizing reactance, the line current, power factor, and input power can be calculated, if desired.

EXAMPLE 8.5

The wound-rotor induction motor of Example 8.1 has a rotor reactance $X_R = 0.352 \ \Omega$ and a rotor resistance $R_R = 0.088 \ \Omega$.

a. Determine the torque-speed characteristic for this machine.

b. Repeat (a) with various additional rotor resistances R inserted, of values 0.1 Ω, 0.264 Ω ($R + R_R = X_{BR}$), and 0.8 Ω.

Solution

a. From Example 8.1 the following data apply:

$$E_{BR} = 88 \text{ V} \qquad n_s = 1800 \text{ r/min}$$

Because of the repeated calculations this exercise was referred to the following computer program.

```
C       Program calculates torque-speed characteristic of an I.M.
C
        REAL Ns,Ir,Nr
        Ns=1800.
        PI=3.14.
        Ebr=88.
        Rr=0.088
        Xbr=0.352
        J=1
10      WRITE(2,11) Rr
11      FORMAT(/////10X,4H Rr= ,F5.3/)
        WRITE(2,4)
4       FORMAT(6X,5H slip,4X,12H Rotor Speed,5x,13H Shaft Torque/)
        DO 3 I=1,20
        s=I
        s=s/20.
        Ir=Ebr/SQRT((Rr/s)*(Rr/s)+Xbr*Xbr)
        RPI=3.*Ir*Ir*Rr/s
        RPD=RPI*(1.-s)
        Nr=Ns*(1.-s)
```

```
        Wr=2.*PI*Nr/60.
        IF(s.GE.1.) GO TO 5
        T=RPD/Wr
        GOTO 6
    5   T=RPI/(2.*PI*Ns/60.)
    6   WRITE(2,7) s,Nr,T
    7   FORMAT(5X,F5,2,3X,F11,0,3X,F12.2)
    8   CONTINUE
        J=J+1
        IF(J.EQ.5) STOP
        IF(J-3)12,13,14
   12   Rr=0.188
        GO TO 10
   13   Rr=Xbr
        GO TO 10
   14   Rr=0.888
        GO TO 10
        STOP
        END
```

b. The results calculated are represented graphically in Figure 8.13. It is apparent that for increased rotor resistances the slip at which maximum torque occurs moves toward the origin. With the total rotor circuit resistance equal to the rotor reactance $R + R_R = X_{BR}$, the startup torque equals the maximum torque. Further increasing R gives curve 4 in Figure 8.13. As can be seen, this characteristic has a negative slope throughout the operating range. This negative slope represents positive damping in control systems applications, which is desirable from a stability point of view. Two-phase induction motors having such very high rotor resistance designs are usually referred to as _servomotors_.

8.11   Determination of Equivalent Circuit Parameters

To use the equivalent circuit model developed for the induction motor in the calculation of performance characteristics, values for the circuit parameters must be assigned. For this, measurements are required from tests, of which the no-load test and the blocked-rotor test are the main evaluations to be carried out. These tests are comparable, respectively, to the open-circuit and short-circuit tests on a transformer.

No-Load Test

As implied, the motor is run with the load uncoupled and rated voltage supplied to the stator. In case of a wound-rotor induction motor, the rotor terminals are shorted out. Measure

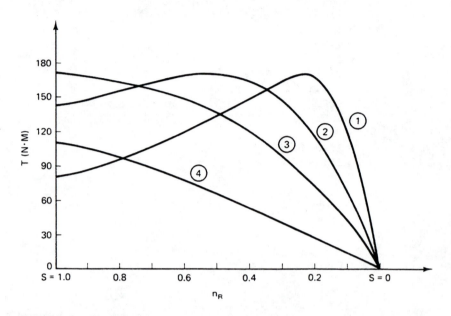

FIGURE 8.13. Representation of calculated data of Example 8.6 using the computer program for (1) $R_r = 0.088 \ \Omega$; (2) $R_r + R = 0.188 \ \Omega$; (3) $R_r + R = X_{BR} \Omega$; (4) $R_r + R = 0.888 \ \Omega$.

$$V_{NL} = \text{line-line stator voltage}$$
$$I_{NL} = \text{line current}$$
$$P_{NL} = \text{three-phase power input}$$

Since the motor at no load runs at a very low value of slip, the no-load rotor copper loss is negligible. Therefore, the input power consists of core loss P_c, friction and windage loss P_{fr+w}, and stator copper loss:

$$P_{NL} = P_c + P_{fr+w} + 3I_{NL}^2 R_s \qquad (8.23)$$

This permits the sum of the friction, windage, and core loss to be evaluated (giving a total loss normally referred to as the *rotational loss* P_{rot}):

$$P_{rot} = P_{NL} - 3I_{NL}^2 R_s \qquad (8.24)$$

where the stator resistance per phase R_s is obtained from a resistance measurement at the stator terminals. If the stator winding is assumed to be Y-connected, the value of R_s per phase is half that of the resistance value measured between terminals. The power factor under no-load conditions is low, so that the circuit behaves essentially reactive. The input current is at least 30% of rated value, depending on the size motor. These facts suggest that the magnitude of R_s is small compared to X_m. Also, the resistance element R_{rot},

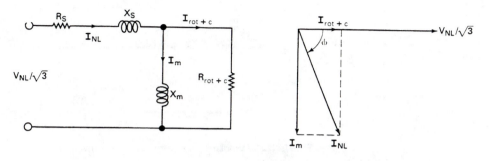

FIGURE 8.14.   Induction motor no-load test.

representing the rotational losses as indicated in Figure 8.14, must then be large compared to X_m. Furthermore, for conventional induction machines, $X_m \gg X_s$. This implies that under the stated assumptions the input impedance at no load is approximately

$$X_m = \frac{V_{NL}}{\sqrt{3}\, I_{NL}} \qquad \leftarrow \text{wrong} \tag{8.25}$$

which is the value of the magnetizing reactance.

Blocked Rotor Test

As the name implies, the blocked rotor test is performed with the rotor blocked, such that the rotor is prevented from turning. Since it cannot turn, $n_R = 0$ and the slip $s = 1.0$ or 100%. This corresponds to the condition at startup and we would expect currents that are five to seven times their rated value. It is for this reason, as with the transformer during the short-circuit test, that the applied stator voltage is reduced to such a voltage, permitting rated stator current to flow. Furthermore, at this greatly reduced input voltage, about 10 to 20% of rated voltage, the air gap flux is relatively small, implying that X_m is much larger than normal. Therefore, the magnetizing reactance is neglected and the equivalent circuit with $R'_R/s = R'_R$ reduces to that of Figure 8.15.

During this test, measure

$$V_{BR} = \text{line-to-line stator voltage}$$
$$I_{BR} = \text{line current}$$
$$P_{BR} = \text{three-phase power input}$$

Assuming a Y connection,

$$Z_e = \frac{V_{BR}}{\sqrt{3}I_{BR}} = \sqrt{(R_s + R'_R) + j(X_s + X'_R)} \tag{8.26}$$

$$R_e = \frac{P_{BR}}{3I_{BR}^2} = R_s + R'_R \tag{8.27}$$

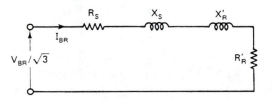

FIGURE 8.15.   Induction motor block-rotor test.

and

$$X_e = \sqrt{Z_e^2 - R_e^2} = X_s + X_R'$$ (8.28)

where Z_e, R_e, and X_e are equivalent motor impedance, resistance, and reactance values per phase, respectively, in stator terms.

   Since R_s was measured separately, the rotor resistance in stator terms is

$$R_R' = R_e - R_s$$ (8.29)

The division between X_s and X_R' is relatively unimportant, since this cannot be determined on a squirrel-cage motor. For wound-rotor induction machines it is generally accepted to assume

$$X_s = X_R' = 0.5 X_e$$ (8.30)

When a wound-rotor induction motor is used, it is generally employed with added external resistances in the rotor circuit for the purpose of starting and speed control. For this application it is necessary to determine the effective turns ratio between rotor and stator, in order to arrive at the actual rotor resistance. The rotor resistance R_R, of course, is measured on the slip rings in a fashion similar to how R_s is measured on the stator side.

   The turns ratio is readily determined by measuring the voltage at the slip rings, after having opened the rotor circuit. The rotor does not turn during this test since no rotor currents can flow. The motor then behaves as a transformer with the secondary open-circuited. If the measured voltage between slip rings is V_{sr} with rated voltage V_L applied to the stator, the transformation ratio is

$$a = \frac{V_L}{V_{sr}} = \frac{N_s}{N_r}$$ (8.31)

where N_s is the stator turns per phase and N_r is the rotor turns per phase. In this equation it is assumed that the rotor is also Y-connected (as it normally is), so that the ratio of the line voltage to that of the voltage measured between slip rings is identical to that of their corresponding phase voltages. The actual rotor reactance is then

$$X_R = \frac{X_R'}{a^2}$$ (8.32)

Similarly, the actual rotor resistance when obtained from Eq. 8.29, namely

$$R_R = \frac{R'_R}{a^2} \qquad (8.33)$$

should agree with that measured from the dc resistance measurement, which may have to be corrected to obtain the effective ac resistance.

EXAMPLE 8.6

A 5-hp 220-V 60-Hz four-pole three-phase induction motor was tested and the following data were obtained.

No-load test: $V_{NL} = 220$ V, $P_{NL} = 340$ W, $I_{NL} = 6.2$ A

Blocked-rotor test: $V_{BR} = 49.4$ V, $P_{BR} = 360$ W, $I_{BR} = 13.9$ A

The dc resistance measurement on the stator winding gives a 4.0-V drop between terminals, when dc current flows equal to the rated current (i.e., 13.9 A). Calculate the efficiency of the motor when operating at a slip of 0.04.

Solution   From the dc resistance test (V-A method),

$$R_{dc} = \frac{4}{2 \times 13.9} = 0.144 \ \Omega/\text{phase}$$

assuming a Y connection.

The effective ac resistance is taken to be 1.25 times the dc value, consistent with earlier work. Thus

$$R_s = 1.25 \times 0.144 = 0.18 \ \Omega$$

From the no-load test,

$$X_m = \frac{220}{\sqrt{3} \times 6.2} = 20.5 \ \Omega$$

$$P_{rot} = 340 - 3 \times 6.2^2 \times 0.18 = 319 \ \text{W}$$

From the blocked rotor test,

$$R_e = \frac{360}{3 \times 13.9^2} = 0.62 \ \Omega$$

$$R'_R = 0.62 - 0.18 = 0.44 \ \Omega$$

$$Z_e = \frac{49.4}{\sqrt{3} \times 13.9} = 2.05 \ \Omega$$

$$X_e = \sqrt{2.05^2 - 0.62^2} = 1.96 \ \Omega = X_s + X'_R$$

Thus $X_s = 0.98$ Ω and $X'_R = 0.98$ Ω.

The resulting equivalent circuit in stator terms becomes the circuit illustrated in Figure 8.16. The performance for $s = 0.04$ can now be calculated following the procedure as indicated in Example 8.3. Only the essential parameters are calculated here to arrive at the efficiency, η.

$$\frac{R'_R}{s} = \frac{0.44}{0.04} = 11.0 \ \Omega$$

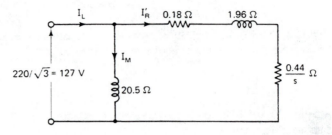

FIGURE 8.16.   Equivalent circuit for Example 8.6.

$$I'_R = \frac{127}{0.18 + 11.0 + j1.96} = 11.2\angle - 9.9° = 11 - j1.93 \text{ A}$$

$$I_m = \frac{127}{j20.5} = -j6.2 \text{ A}$$

The line current

$$I_L = 11.0 - j(1.93 + 6.2) = 13.7\angle - 36.5° \text{ A}$$

Therefore,

$$\text{RPI} = 3 \times 11.2^2 \times 11.0 = 4140 \text{ W}$$
$$\text{RPD} = 4140 \times (1 - 0.04) = 3974 \text{ W}$$
$$\text{RCL} = 0.04 \times 4140 = 166 \text{ W}$$
$$P_{out} = 3974 - 319 = 3655 \text{ W}$$

and the total losses are

$$319 + (3 \times 11.2^2 \times 0.18) + 166 = 553 \text{ W}$$

Thus the efficiency

$$\eta = \frac{3655}{3655 + 553} \times 100 = 86.9\%$$

As Example 8.6 indicates, the performance data for the machine can be determined by performing relatively simple tests, to establish the equivalent circuit parameters. With regard to the efficiency, as explained before, it can be determined by measuring the input and output powers directly. In the event of large machines, it may become impossible to simulate an actual load, and therefore it is more expedient to resort to measuring the losses as indicated, to arrive at the motor efficiency.

8.12   *Selecting an Induction Motor*

As discussed previously, the NEMA class B induction motor is industry's general-purpose motor. It has good starting properties and is suitable for ventilating equipment such as fans and blowers, for pumps of the centrifugal type, and so on. If the torque-speed curve of a centrifugal pump is superimposed on the curve for the induction motor, as shown in Figure 8.17, it can be seen that the load is easily started and there is ample accelerating torque. Thus full speed is quickly reached. If this motor were to drive a load that requires a high starting torque, but only requires a relatively small running torque, such as a loaded conveyor belt, the motor might not be able to start the belt, as Figure 8.18 shows. Selection of a larger motor may not be economical. This suggests selection of a motor with higher starting-torque capabilities, such as a class C induction motor. Since the class D motor would also provide the required higher starting torque, it would not be a proper selection since it runs at a higher slip and therefore at lower efficiency. Even though the class C motor is the clear choice here, it must still be checked to make sure that the motor meets the startup torque requirement if voltage drops are prevalent in the supply system. This is important since the net torque developed by an induction motor is proportional to the square of the applied voltage, as is apparent from Eq. 8.22. The rationale for this is as follows. The torque developed is proportional to the flux and the current, but the flux itself is proportional to the voltage. Therefore, the developed torque $T \propto V^2$. To illustrate this dependency, the starting torque of the motor in Example 8.6 will be determined for full-voltage starting, considering the following assumption.

At the instant of starting, the rotor is not turning, and the friction and windage loss is therefore zero. The iron loss at starting is greater than normal, however, because the rotor frequency equals the stator frequency. It seems reasonable, therefore, to assume that the increased iron loss caused by the rotor compensates for the decreased rotational loss. It

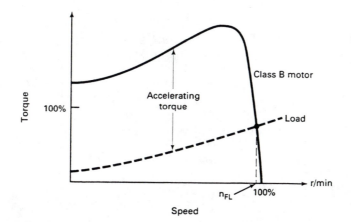

FIGURE 8.17.   Class B induction motor and load torque-speed character-istic.

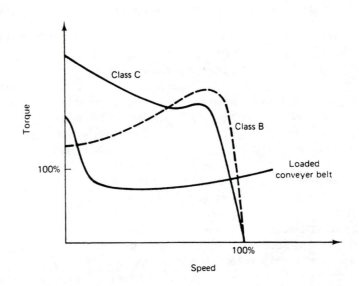

FIGURE 8.18.   High starting torque of class C motor.

is under this condition that the rotational loss may be considered constant throughout the normal operating range of the machine.

EXAMPLE 8.7

Calculate the starting torque of the motor in Example 8.6 when started at full voltage.

Solution   From Example 8.6, the rotational loss = 319 W. The line current at 220 V, with the rotor blocked, is

$$\frac{220}{49.4} \times 13.9 = 61.9 \text{ A}$$

The power input at 220 V, rotor blocked, is

$$\left(\frac{220}{49.4}\right)^2 \times 360 = 7140 \text{ W}$$

The stator copper loss is

$$P_{stator} \times (61.9)^2 \times 0.18 = 2071 \text{ W}$$

and

$$\text{RPI} = 7140 - 2071 - 319 = 4750 \text{ W}$$

Therefore,

$$T_{st} = \frac{4750}{2\pi \times 1800/60} = 25.2 \text{ N·m}$$

8.13 Starting an Induction Motor

As we have seen in the various examples, the starting currents of large induction motors can be excessive. Most modern induction motors are designed to withstand the mechanical forces and inrush currents associated with full-voltage starting. Although any size of motor may be started this way, it must be recognized that objectionable line-voltage fluctuations (called *voltage flicker*) generally occur. Because of this, and the consequent effect on the distribution system, power utilities have set rules and regulations regarding the size motors that may be started directly on the line.

To limit the objectionable effects of across-the-line starting, various types of reduced-voltage starters have been produced to limit the starting current. But since the starting torque is proportional to the applied voltage squared, care must be exercised when starting under reduced voltage, that the starting torque is adequate to accelerate the motor and load. Even a modest reduction to 80% of rated voltage will reduce the starting torque to 64% of its nominal value. It should be kept in mind that the problem of providing the proper starting equipment is not always one of merely selecting a starter that satisfies the rating of the motor. The motor itself must be selected to meet the requirements of the particular application.

Although numerous methods are employed, only some of the generally used starting methods will be discussed in relation to the squirrel-cage induction motor. Wound-rotor induction motors are started by inserting resistors in the rotor circuits.

Full-Voltage Starting Method

A simple full-voltage automatic starter for an induction motor is shown in Figure 8.19a. Note that the essential components are shown in the various starting circuits to be presented but not the necessary protection circuitry, such as circuit breakers, fuses, and main switches.

The motor is started by pressing the momentary-contact start button. When the start button is pressed, the control circuit is energized and causes the main starter relay M to energize. Its auxiliary contacts M close, thereby starting the induction motor, IM. When the start button is released its contact opens, but the starter auxiliary contact M, which is wired in parallel with the start button, continues to maintain the circuit, and as such is called the *holding circuit*. The motor would be disconnected from the line by pressing the stop button, since the contactor M would then be deenergized.

In the event of a power interruption or greatly reduced voltage, the contactor coil M will be deenergized, thereby opening the holding circuit. When voltage is restored, the motor cannot restart automatically. The start button must be pressed again. This safety feature is called *low-voltage protection*, which is one of the important advantages of magnetic control. Flexible control is another feature of the magnetic starters. A pushbutton start-stop switch box for energizing the main contactor coil can be located any distance from the machinery being driven. Furthermore, several such pushbutton stations can be used to provide multiple-location operation of the same starter and motor. Instead of a pushbutton arrangement, a relay, timer, or pressure switch, for instance, could be used to provide automatic operation in addition to remote control.

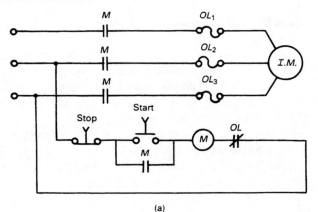

(a)

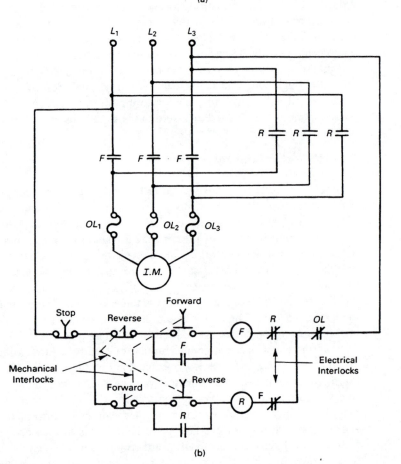

(b)

FIGURE 8.19. Diagram of full-voltage automatic starters for three-phase squirrel-cage induction motors: (a) simple starter circuit; (b) circuit includes reverse-run.

One of the basic characteristics of squirrel-cage motors is that the direction of rotation can be changed by reversing any two of the incoming line-voltage leads. Reversing motor rotation can also be provided with a standard magnetic contactor with the proper electrical and mechanical interlocking. Figure 8.19b shows how it is done. The interlocking feature prevents both sets of contacts, forward and reverse run, from being closed at the same time, even momentarily. There is no possibility of a short circuit then.

When the forward-run button is pressed in Figure 8.19b, contactor F is energized through the stop button, upper contacts of the reverse-run button, and normally closed contact R of the reverse-run contactor, and the normally closed overload contact OL. The three main contacts of the forward contactor close to start the motor. Coil F is kept energized by its holding contact once the start button is released.

As seen, another normally closed contact F opens the reverse-run circuit to ensure that the forward-run and reverse-run circuits are electrically interlocked. The starter, apart from being interlocked electrically, is mechanically interlocked as well, as indicated by the dotted line in the diagram. This prevents a short circuit if both forward-run and reverse-run pushbuttons are pressed simultaneously. In addition, overload protection is provided by thermal elements OL placed in the motor leads. Should the motor overheat, these normally closed contacts are opened, thereby deenergizing the contactor, which in turn disconnects the motor from the line. If the reverse-run button in Figure 8.19b is pressed, the forward-run circuit opens through the normally closed contact of the reverse-run pushbutton. This deenergizes the forward-run contactor coil, opening the forward contacts in the motor circuit. The normally closed auxiliary contact F in the reverse-run circuit is thereby closed. The reversing circuit is thus completed through the momentarily closed lower contacts of the reverse button, the lower contacts of the forward-run button, and auxiliary contact F. Reverse-run contact R is now energized, and the reverse-run contacts in the motor circuit are closed. Auxiliary contact R opens in the forward-run circuit for added safety. The reverse-run contactor is kept energized by holding contact R across the reverse button, until the forward-run button or stop button is pressed. Note that lines L_1 and L_3 are interchanged in the reverse-run position as compared to the forward-run mode.

Y-Δ *Starting Method*

For this method of starting, the end leads of each phase winding must be brought out to the terminal box. The reduced voltage for starting is obtained by connecting the winding in star. This means that the phase winding voltage will only be $1/\sqrt{3}$, or 58% of its rated value. When the motor reaches 75 to 80% of rated speed, full-rated voltage is applied to the windings by reconnecting them in delta. The starting torque is reduced to one-third of normal value [since $T \propto V^2$, $T_{st} = 1/(\sqrt{3})^2 = 0.333\times$ rated value]. If this starting torque is sufficient for the particular application, this method should be given due consideration because of its simplicity and therefore, lower cost.

As shown in Figure 8.20, the operation is as follows: when the start button is pressed, the main contactor M energizes, closing its contacts, thereby connecting winding terminals T_1, T_2, and T_3 to the line. Also, contactor S energizes through the normally closed contact TO of the timing relay TR, thereby closing contacts S to complete the star-point connection.

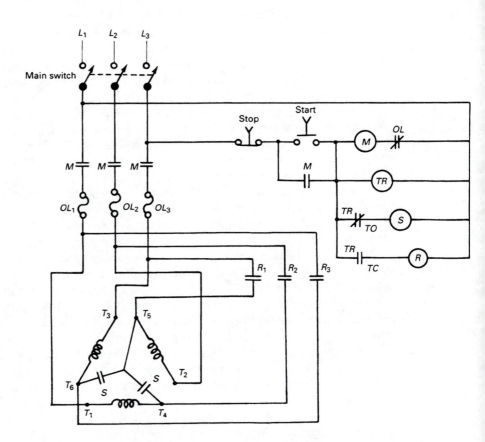

FIGURE 8.20.   **Automatic Y-Δ starter.**

After timing relay TR times out, the S contactor deenergizes first, which is followed by the closing of the R contacts. As is readily seen, terminals $T_1 - T_6$, $T_2 - T_4$, and $T_3 - T_5$ are joined, thereby connecting the winding in delta.

Autotransformer Method

Figure 8.21 shows a starting method using a three-phase autotransformer to reduce the line voltage at startup. The autotransformer may have a single tap or multiple taps, depending on the application. When multiple taps are employed they are usually made at the 50, 65, and 80% points, so that selection of the proper tap can be made for the proper starting torque. The corresponding starting torques will be 25, 42, or 64% of rated torque, respectively. At the same time the starting current is reduced in proportion to the impressed voltage, and reduced further by the transformer action. Hence the starting current is decreased in proportion to the impressed voltage squared, or to 25, 42, or 64% of full-voltage starting.

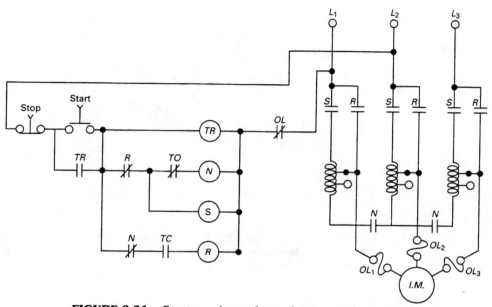

FIGURE 8.21.   Starter using a three-phase autotransformer.

The starting sequence is as follows. Pressing the start button causes contactors S and N to be energized, closing their respective contacts and start timer relay TR. The N contacts in conjunction with the S contacts connect the primaries of the autotransformer in a Y connection to the supply voltage. The motor will start at a reduced supply voltage depending on the tap settings. After timing relay TR times out, the thermally operated contact TO opens, dropping out contactor N and closing TC, thereby energizing contactor R. Contacts R put the induction motor directly on the line and deenergizes contactor S. The autotransformer is thereby disconnected from the line.

As you can now appreciate, all starting methods somehow reduce the line voltage (and consequently the line current) to the motor at startup. How this is accomplished characterizes the particular starting method. Other less costly but also less effective ways of reducing the starting voltage may be accomplished by connecting starting resistors or reactors in series with the motor during starting. When the motor has accelerated to about 75% of rated full-load speed, the resistors or reactors are then shorted out, thereby connecting the motor across the line. The reactor method of starting has an advantage over the resistor starting method in that the available acceleration torque is larger. Figure 8.22 indicates this. The reason for this is attributable to the reactor voltage being out of phase with the motor voltage. The overall effect is a larger resultant voltage for the reactor system than for the resistor system. This is most desirable, especially for accelerating loads for which the torque increases with speed, such as compressor and fan loads.

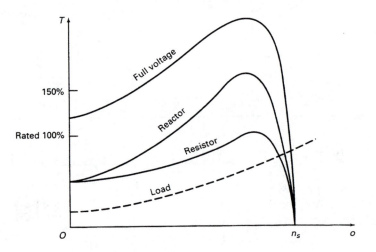

FIGURE 8.22. Accelerating torque characteristics for starting methods indicated of starting squirrel-cage induction motors for an assumed load torque.

Wound-Rotor Method

Secondary resistor starters, as opposed to primary resistor starters just discussed, are used for the acceleration of wound-rotor motors. They consist of an across-the-line starter and one or more accelerating contacts to shunt resistances in the secondary rotor circuit. The resistances connected to the slip rings of the motor are meant for starting duty only, and this starter cannot be used for speed regulation. The operation of the accelerating contactors is controlled by a timing relay which provides timed acceleration and eliminates the danger of improper startup procedures. A typical starter diagram with four points of acceleration is shown in Figure 8.23. As can be seen, balanced rotor-resistance conditions are maintained by the closing of successive sets of contacts A_1, A_2, and A_3. Since the final step comprises the full-speed condition, short-circuiting three pairs of lines helps to reduce contact resistances to zero. Figure 8.24 shows the torque-speed characteristic with the three different external resistances added to the rotor circuit. During startup the resistances are eliminated much in the fashion of a dc motor starter. By proper selection of the starting resistors, any torque requirement can be met within the capabilities of the motor.

8.14 Induction Motor Speed Control

The speed of a squirrel-cage induction motor operating from a constant voltage and frequency supply, is essentially constant. The motor normally runs at a slip of about 2 to 5% of synchronous speed. When operating on the power system the frequency is fixed, the

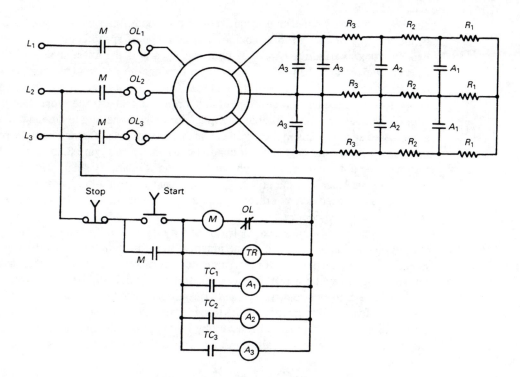

FIGURE 8.23. Typical four-point controller.

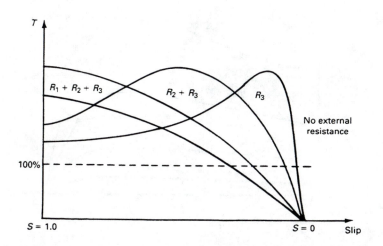

FIGURE 8.24. Starting of wound-rotor with four-point starter.

slip is determined by the load torque, and speed control can be achieved by adjustment of the voltage. Although this provides a wide range of speed control, the ohmic losses become excessive and the efficiency poor, particularly at the lower speeds. This is so because at a given slip the motor current is proportional to the voltage, whereas the torque developed varies as the square of the voltage. Therefore, at reduced speeds an appreciable larger current is required to develop a large torque. It suggests therefore, that the voltage control methods are more suitable in applications where the torque demand is also lower at increased speeds. This in fact is true in such applications as fan and pump drives, where the load characteristic is somewhat quadratic, as shown in Figure 8.25. As is shown, the torque demand for starting and high-slip operation is relatively small. However, although performing satisfactorily the efficiency is poor, and generally the motor is derated at low speeds to avoid excessive overheating due to reduced ventilation. This method of control is referred to as *slip control*.

Another method of speed control is to vary the supply frequency to the motor. Although this is simple in principle, the practical implementation is generally not so convenient. With the advent of power electronic devices many sophisticated variable-frequency inverter systems are now available as a "shelf item," which will be discussed in principle, shortly. In such an electronic drive, the frequency is reduced for starting. Figure 8.26 shows the effect of variable frequency on the torque-speed characteristic and the load speed. The maximum torque is maintained constant by keeping the volts/frequency ratio the same so that the air-gap flux is essentially constant. This follows from Eq. 7.6, namely

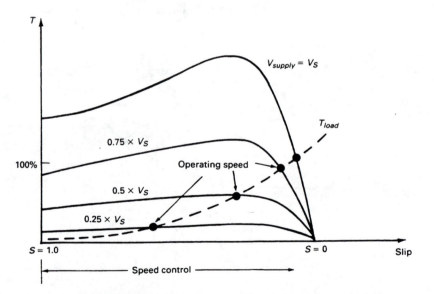

FIGURE 8.25. Speed control by motor voltage variation.

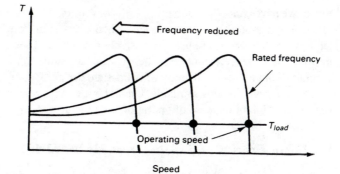

FIGURE 8.26.   Speed control of induction motor by varying the supply frequency, keeping the air-gap flux constant.

$$\frac{V}{f} = 4.44 \, N \, \phi \qquad (8.34)$$

As indicated, this type of control is suitable for driving a constant-torque load.

If the frequency is varied while the stator applied voltage remains constant, the air-gap flux and stalling torque would decrease with frequency, as depicted in Figure 8.27. This type of speed control would be more suitable for traction applications where larger torques are needed at startup but smaller torques at running speeds. A typical solid-state drive system that may accomplish this will be described in principle in the following section.

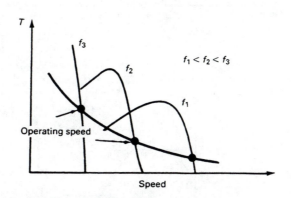

FIGURE 8.27.   Speed control of an induction motor by varying the supply frequency and constant terminal voltage.

8.15   *Solid-State Drives*

So far, we have concerned ourselves with conventional starting and speed control techniques of induction motors. As was evident when discussing dc motor speed control in Chapter 4, solid state drives are used extensively nowadays. Therefore, we will want to examine in principle how we can control the speed of an induction motor using modern controls, without doing an in-depth study of solid-state drives. Depending on the application, there are several drive systems available, but all have certain considerations in common, namely: initial cost, reliability, ease of operation, and maintenance.

Variable-Frequency Induction Motor Drive

As indicated by Eq. 8.1, the synchronous speed of an induction motor can be altered by varying the frequency. However, from Eq. 8.34 it is seen that when the frequency is decreased, the magnetic flux increases (assuming linear behavior). If currents are to remain at their normal values and saturation of the magnetic circuit is to be avoided, the applied voltage must also be varied proportionally to the frequency, such that the ratio V/f remains constant. This type of control is generally termed *constant volts per hertz*. One of many possible circuits to control the speed of an induction motor is schematically represented in Figure 8.28a. The motor-voltage (V_m) control is obtained by controling the thyristor gating of a three-phase controlled rectifier (ac to dc), while the motor frequency is changed by the inverter circuir (dc to ac). The operation of these circuits will not be discussed here.

The resultant motor line-to-line voltage is a quasi square wave of 120° pulse width. However, because of the motor inductance the motor current is essentially sinusoidal. as

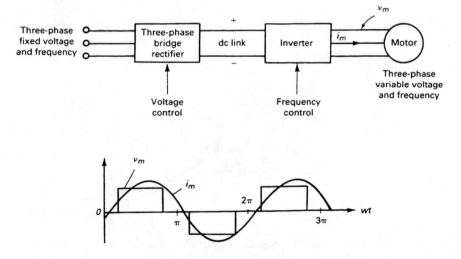

FIGURE 8.28.   Speed control system for induction motor.

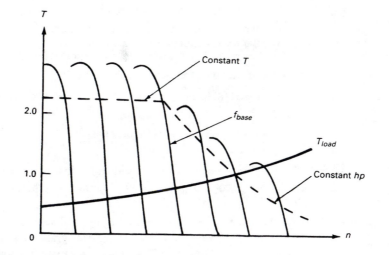

FIGURE 8.29. **Torque-speed characteristics of an induction motor variable-voltage, variable-frequency control system.**

indicated in Figure 8.28b. The torque-speed characteristic for this method of control is shown in Figure 8.29. At the base frequency f_{base} the machine terminal voltage is at rated value. Below this frequency, the air gap flux is maintained constant by decreasing V_m with frequency such that V/f is constant; hence the same maximum torque is available. At frequencies higher than f_{base}, since V_m should not exceed rated value, the air gap flux decreases and so does the maximum available torque. This corresponds to the field weakening control scheme used with dc motors. Constant horsepower operation is possible in the field-weakening region.

PROBLEMS

8.1 At what speed will a 12-pole 60-Hz induction motor operate if the slip is 0.06?

8.2 A 60-Hz induction motor runs at 860 r/min at full load. Determine

    a. the synchronous speed,

    b. the frequency of the rotor currents, and

    c. the rotor speed relative to the revolving field.

8.3 A three-phase induction motor runs at almost 1200 r/min at no load and 1140 r/min at full load when supplied with power from a 60-Hz three-phase line.

    a. How many poles has the motor?

    b. What is the percent slip at full load?

    c. What is the corresponding frequency of the rotor voltages?

    d. What is the corresponding speed of

        i. the rotor field with respect to the rotor?
        ii. the rotor with respect to the stator?
        iii. the rotor field with respect to the stator field?

    e. What is the rotor frequency at the slip of 10%?

8.4 A three-phase 60-Hz induction motor runs at 594 r/min at no load and at 558 r/min at full load.

    a. How many poles does this motor have?
    b. What is the slip at rated load?

8.5 If the EMF in the stator of a six-pole induction motor has a frequency of 60 Hz, and that in the rotor of 2 Hz, at what speed is the motor running and what is the slip?

8.6 If the phase sequence of the currents in Figure 8.3a is reversed, for example, by reversing i_b and i_c (*acb* sequence), demonstrate graphically, using the convention in the text as depicted in Figure 8.3b, that the resultant flux rotates counterclockwise.

8.7 A three-phase 60-Hz six-pole 220-V wound-rotor induction motor has its stator connected in delta and its rotor in star. There are 80% as many rotor conductors as stator conductors. For a slip of 0.04, calculate the rotor-induced voltage per phase.

8.8 A three-phase 60-Hz six-pole 220-V wound-rotor induction motor has its stator connected in Δ and its rotor in Y. The rotor has half as many turns as the stator. Calculate the rotor-induced voltage and its frequency if normal voltage is applied to the stator and

    a. the rotor is at rest,
    b. the rotor slip is 0.04, and
    c. the rotor is driven by another machine at 800 r/min in the direction opposite to that of the revolving field.

8.9 For slip values of $s = 0.05$ to 1.0 in suitable steps, calculate the performance characteristics for the motor in Example 8.2. Plot efficiency, line current, and power factor as a function of developed horsepower.

8.10 A three-phase 20-hp 220-V 60-Hz six-pole Y-connected squirrel-cage induction motor has the following parameters per phase in stator terms: $R_s = 0.126$ Ω, $R'_R = 0.094$ Ω, $X'_e = 0.46$ Ω, $X_m = 9.8$ Ω. The rotational losses are 560 W. For a slip of 3%, find

    a. the line current and power factor,
    b. the horsepower output and shaft torque, and
    c. the efficiency.

8.11 An induction motor has an efficiency of 0.88 when the load is 25 hp. At this load, the stator copper and rotor copper loss each equal the iron loss. The mechanical losses are one-third of the no-load loss. Calculate the slip.

8.12 A dc test is performed on a 440-V Δ-connected 60 Hz induction motor. If $V_{dc} = 20$ V and $I_{dc} = 28.8$ A, what is the effective stator resistance R_s/phase? Assume a factor of 1.20 accounts for the skin effect, etc.

8.13 A three-phase 125-hp 440-V 60-Hz eight-pole Y-connected induction motor has the following electric circuit parameters on a per phase basis referred to the stator:

$$R_s = 0.068 \ \Omega \quad X_s = X'_R = 0.224 \ \Omega$$
$$R'_R = 0.052 \ \Omega \quad X_m = 7.68 \ \Omega$$

The rotational losses are 2400 W. Determine for a slip of 3%

a. the line current and power factor,
b. the output horsepower and torque, and
c. the efficiency.

8.14 For the machine of Problem 8.13, calculate

a. the slip at which maximum torque occurs,
b. the speed at which this occurs, and
c. the line current under this condition.

8.15 For the machine of Problem 8.13, determine the starting torque and the value of the starting current.

8.16 The nameplate of a squirrel-cage induction motor has the following information: 25 hp, 220 V, three-phase, 60 Hz, 830 r/min, 64 A per line. If the motor takes 20.8 kW when operating at full load, calculate

a. the slip,
b. the power factor,
c. the torque, and
d. the efficiency.

8.17 A 7.5-hp 220-V four-pole three-phase induction motor was tested and the following data were recorded:
No-load test: $V_{NL} = 220$ V, $P_t = 320$ W, $I = 6.4$ A
Blocked-rotor test: $V_{BR} = 46$ V, $P_t = 605$ W, $I = 18$ A
The effective ac resistance between stator terminals is 0.64 Ω, and the slip at full load is 4%. Determine

a. the equivalent electrical circuit on a per phase basis for this motor,
b. the input current and power factor when delivering full load, and
c. the efficiency and motor speed in part (b).

8.18 A 220-V four-pole 10-hp 60-Hz Y-connected three-phase induction motor develops its full-load air-gap torque at 3.6% slip when operating at 60 Hz and 208 V. The per phase circuit model parameters for the motor are

$$R_1 = 0.34 \ \Omega \quad X_m = 15.2 \ \Omega$$
$$X_s = 0.46 \ \Omega \quad X'_r = 0.46 \ \Omega$$

The rotational losses may be neglected in this problem. Determine

   a. the value of the rotor resistance R'_r,

   b. the maximum torque and the slip value at which it occurs, and

   c. the starting torque if started at rated voltage.

8.19 For the motor in Problem 8.17, calculate the starting torque when it is started at 127 V.

8.20 A 25-hp 440-V squirrel-cage induction motor has a starting torque of 112 N·m and a full-load torque of 83 N·m. The starting current of the motor is 128 A when rated voltage is applied. Determine

   a. the starting torque when the line voltage is reduced to 300 V,

   b. the voltage that must be applied in order to develop a starting torque equal to the full-load torque,

   c. the starting current when the voltage is reduced to 300 V, and

   d. the voltage that must be applied at startup in order not to exceed the rated line current of 32 A.

8.21 Calculate the relative values of the starting torque and the starting current of a three-phase squirrel-cage induction motor when started by

   a. rated voltage,

   b. a star-delta starter, and

   c. a three-phase autotransformer having 60% voltage taps.

9

Single-Phase Motors and Special Machines

In this chapter we discuss single-phase fractional horsepower motors and a variety of small machines that are used for special and general purpose applications. These machines have either unusual operating characteristics or construction details not found on the common type of machines discussed so far. For example, permanent magnet fields, brushless dc motors, stepper motors, printed circuit board armatures, and so on. Many of these devices are employed together with electronic circuitry to provide precise speed control and positioning control. We will present unique features for the most common variety of small electric motors.

9.1 Fractional-Horsepower Machines

Production of Torque

In the three-phase induction motor, the three-phase distributed armature winding sets up a rotating magnetic field which is fairly constant in magnitude and rotates at synchronous speed. In single-phase induction motors we have only a simple field winding excited with alternating current; therefore, it does not have a true revolving field, and hence it is not inherently self-starting. Various methods have been devised to initiate rotation of the squirrel cage, and the particular method employed to start the motor will designate the specific type.

First let us examine the behavior of the magnetic field as set up by an ac current in the single-phase field winding. With reference to Figure 9.1, we have current flowing in the field winding. If this current is sinusoidal, then, neglecting saturation effects of the magnetic iron circuit, the flux through the armature will vary sinusoidally with time. The magnetic field created is as shown at the particular instant in time; it will reverse during the next half-cycle of the ac supply voltage cycle. Since the flux is pulsating it will induce currents in the rotor bars, which in turn will create a rotor flux which by Lenz's law opposes that of the main field. From this the current direction in the rotor bars can be determined, as shown in Figure 9.1, as well as the torque created between the field and rotor currents. It is apparent that the clockwise torque produced is counteracted by the counterclockwise torque, hence no motion results. Since the field is pulsating, the torque is pulsating, although no net torque is produced over a full cycle of the ac supply frequency.

However, any pulsating field can be resolved into two components, equal in magnitude but oppositely rotating vectors, as shown in Figure 9.2a. The maximum value of the component fields equals one-half of ϕ_{max}. As can be observed, the resultant field of ϕ_1 and ϕ_2 in Figure 9.2a as they rotate at an angular velocity dictated by the supply frequency, must always lie on the vertical axis. The resultant value of these two vectors at any instantaneous time equals the value of the magnetic field as it actually exists. A physical interpretation of the two oppositely rotating field components is as predicted in Figure 9.2b. Each component field glides around the air gap in opposite directions and equal velocities; their instantaneous sum represents the instantaneous resultant field, which changes between $\pm\phi_{max}$. This method of field analysis is commonly known as the *double-revolving field theory*. Each field component acts independently on the rotor and in a fashion similar to that of the

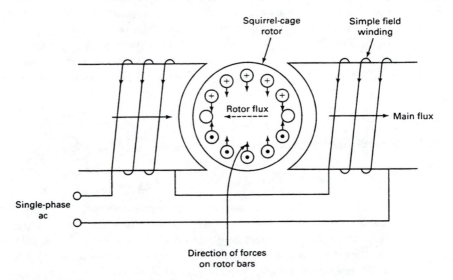

FIGURE 9.1. Torque produced in squirrel cage of single-phase induction motor having a simple field winding.

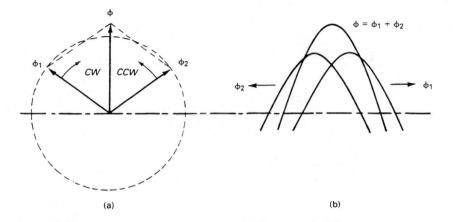

(a)

(b)

FIGURE 9.2.   **Pulsating field resolved into two oppositely rotating fields.**

rotating field in a three-phase induction motor, except that here there are two, one tending to rotate the rotor clockwise, the other tending to rotate it counterclockwise.

Considering the clockwise flux component by itself, it would produce the torque-speed characteristic labeled T_{cw} in Figure 9.3, while the counterclockwise flux component produces the torque T_{ccw}. Observe that at standstill (s = 1.0) the two torque components produced are equal, but directed oppositely. Although the net torque produced at standstill is zero ($T_{cw} = T_{ccw}$), if the rotor were to be advanced in either direction, a net torque will result, and the motor will continue to rotate in the direction in which it has been started. For example, if we assume that the rotor is in some way started in the clockwise direction, the torque T_{cw} will exceed T_{ccw} immediately and the rotor will accelerate in the direction of T_{cw}. Steady-state speed will be reached near synchronous speed at a slip dictated by the load. It is interesting to note that at this slip speed, T_{cw} predominates over T_{ccw}, which is fairly small but exists nevertheless. Also, the rotor operates at a small slip value as far as T_{cw} is concerned, but the slip is nearly 2 with regard to T_{ccw}. This implies induced rotor currents due to T_{ccw} which are at double the line frequency. These rotor currents do not produce any significant countertorque because of their high frequency, since the rotor reactance is many times its value at slip frequency.

In the three-phase induction machine the rotating field strength does not vary appreciably as it rotates, and the field is said to be circular, as shown in Figure 9.4. The field set up by a single-phase motor on the other hand is usually elliptical in shape. To see why this occurs we have to examine the rotor field more closely.

When the rotor is rotating, voltages are induced in the rotor conductors which are in phase with the stator field. Since these voltages are speed dependent they are referred to as *speed EMFs*, as opposed to *transformer EMFs*, which are produced by transformer action. Both are, of course, produced by a changing flux, the speed EMF as a result of relative motion between the field and conductor, the transformer EMF as a result of a pulsating field.

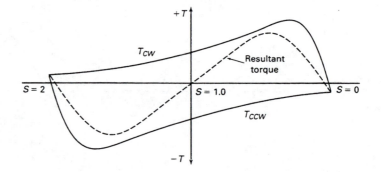

FIGURE 9.3.   **Torque-speed characteristic of a single-phase squirrel-cage induction motor.**

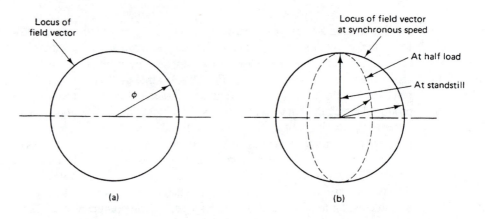

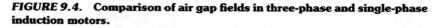

FIGURE 9.4.   **Comparison of air gap fields in three-phase and single-phase induction motors.**

Therefore, since rotor-induced currents flow in the rotor bars, which represent almost entirely a reactive impedance, these rotor currents will lag the rotor-induced voltage by nearly 90°. In turn, the field created by the rotor currents is also displaced by 90° and is known as a *cross field*, as indicated in Figure 9.5. Thus the stator sets up a pulsating field while the rotating rotor sets up a second field which pulsates 90° behind the stator field in space and time. When the rotor rotates at nearly synchronous speed, these two fields will virtually be equal and will combine to produce a nearly circular field. Upon loading the single-phase induction motor its speed will drop, thereby reducing the induced EMF in the rotor. This in turn reduces the cross field and the resulting field becomes elliptical, as shown in Figure 9.4b. Further decreasing the speed until finally, standstill is reached, the resulting field will only pulsate in value along the stator axis and does not rotate.

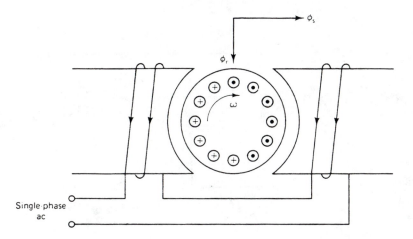

FIGURE 9.5. **Cross field ϕ_r created by rotor rotation.**

Thus, once started, the single-phase motor, having a simple winding as explained, will continue to run in the direction in which it is started. Non-self-starting is not a desirable feature in practice, and modifications are introduced to obtain the torque required to start. To accomplish this a quadrature flux component in time and space with the stator flux must be provided at standstill. Auxiliary windings normally placed on the stator have proved effective in developing this starting torque. The method employed to accomplish this will now be described.

Split-Phase Induction Motors

One of the most widely used types of single-phase motors is the split-phase induction motor. Its service includes a wide variety of applications such as used for refrigerators, washing machines, portable hoists, many small machine tools, blowers, fans, centrifugal pumps, and many others. The essential parts of the split-phase motor are shown in Figure 9.6a. It shows the *auxiliary winding*, also called the *starting winding*, in space quadrature (i.e., 90 electrical degrees displacement) with the main stator winding. The rotor is normally the normal squirrel-cage type. The two stator windings are connected in parallel to the ac supply voltage. A phase displacement between the winding currents is obtained by adjusting the winding impedances, either by inserting a resistor in series with the starting winding or as is generally the practice, by using a smaller-gauge wire for the starting winding. A phase displacement between the currents of 30° can be achieved at the instant of starting. A typical phasor diagram for this motor at startup is illustrated in Figure 9.6b.

When the motor has come up to about 70 to 75% of synchronous speed, the starting winding may be opened by a centrifugal switch, and the motor will continue to operate as a single-phase motor. At the point where the starting winding is disconnected, the motor develops nearly as much torque with the main winding as with both windings connected,

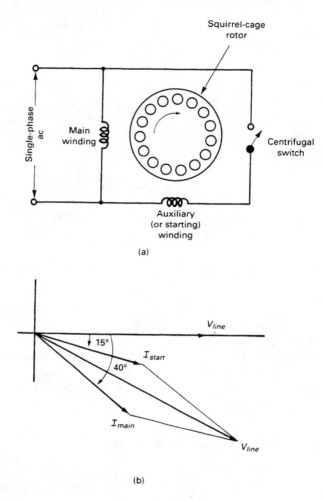

(a)

(b)

FIGURE 9.6.   Split-phase motor: (a) schematic representation; (b) phasor diagram at the instant of starting.

as can be observed from the typical torque-speed characteristic for this type of motor (see Figure 9.7).

The starting winding is designed to take the minimum starting current for the required torque. The locked rotor starting current is typically in the range 5 to 7 times rated current, while the starting torque is about 1.5 to 2 times rated torque. The high starting current as such is not objectionable, since once started it drops off almost instantly. The major disadvantages are the relatively low starting torque and the high slip at which it operates when heavily loaded. As you can appreciate from the earlier discussion, when the speed drops significantly (of course, not to the extent that the centrifugal switch operates), the induced EMF is reduced. This results in an elliptical or pulsating torque, which makes this

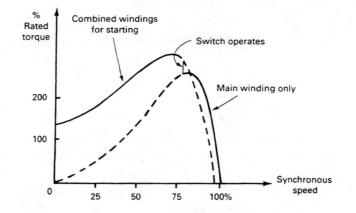

FIGURE 9.7. **Typical torque-speed characteristic of a general purpose split-phase motor.**

motor somewhat noisy. It is precisely for this reason that the split-phase motor is operated and employed where the drive loads themselves are noisy.

Unlike the three-phase induction motor, which may start in either direction, the split-phase motor is factory connected, and as such its direction of rotation is fixed (counterclockwise when viewed from the opposite end of the shaft extension). To reverse the direction of rotation it is necessary to reverse the connection to the starting winding. Again in contrast to its three-phase counterpart, this reversal (plugging) cannot be done under running conditions, since the split-phase motor torque will be much less than the torque developed by the single main winding, and rotation will not reverse.

In the event that the centrifugal switch contacts fuse, the starting winding will be permanently connected in the circuit during normal operation. Although in principle this does not affect motor operation significantly, it must be realized that this winding is designed for intermittent operation (for startup only). Therefore, when permanently connected in the circuit due to switch failure, it will quickly heat up and raise the motor temperature excessively and eventually burn out the windings.

As Figure 9.6b shows, the starting winding current I_s lags the supply voltage by about 15°; the running or main winding current I_{main} lags the voltage by about 40°. Although the currents are not equal, their quadrature or in-phase components with the voltage are nearly the same. This is illustrated in Example 9.1.

EXAMPLE 9.1

A 1/4-hp 120-V split-phase motor draws at the instant of starting a current of 4 A in its starting winding, while the main winding current takes 5.8 A, lagging the supply voltage by 15° and 45°, respectively. At startup, determine:

a. the line current and power factor, and

b. the in-phase components of the currents with the supply voltage.

Solution

a.

$$I_{start} = 4\angle - 15° = 3.86 - j1.04 \text{ A}$$

$$I_{main} = 5.8\angle - 45° = 4.10 - j4.10 \text{ A}$$

$$I_{line} = I_{start} + I_{main} = 7.96 - j5.14 = 9.48\angle - 33° \text{ A}$$

power factor= $\cos(-33°)$ = 0.84 lagging

b. From the calculated results in part (a) we see that the in-phase components of currents with the line voltage, being the real parts of the respective currents, are 3.86 A and 4.10 A for the starting winding and main winding current, respectively. As shown, these components are practically equal at the instant of starting.

Capacitor-Start Motors

In the split-phase motor the phase shift between stator currents was accomplished by adjusting the impedances of the windings by making the starting winding a relatively higher resistance. This resulted in a phase shift of nearly 30°. Since the developed torque of any split-phase motor is proportional to the pole flux produced and the rotor current, it is also dependent on the angle between the winding currents. This implies that if a capacitor is connected in series with the starting winding, the starting torque will increase. This is indeed the case. By proper selection of the capacitor value the current in the starting winding will lead the voltage across it and a greater displacement between winding currents is obtained. This results in a significantly greater starting torque than that obtained in split-phase motors, as Example 9.2 will illustrate. Typical starting torques may be in the range of four times rated torque.

Figure 9.8 shows the capacitor start motor and its corresponding phasor diagram, indicating a typical displacement between winding currents of about 80°. The value of capacitor needed to accomplish this is typically 135 μF for a 1/4-hp motor and 175 μF for a 1/3-hp motor. Since they are rated for ac line voltages their size is about 1 1/2 in. in diameter and 3 1/2 in. long. Contrary to the split-phase motor discussed, the capacitor-start motor under running conditions is reversible. If temporarily disconnected from the supply line, its speed will drop, allowing the centrifugal switch to close. The lead connections to the starting winding are reversed during this interval and the motor reconnected to the supply once the centrifugal switch closes. The resulting rotating field will now rotate opposite to the direction the motor rotates.

Since the current displacement between windings is much larger in this motor than in the split-phase motor, the torque being proportional to this will be much larger and exceed the torque produced by the rotor. Therefore, the motor will slow down, stop, and then reverse its direction. Once up to about 75 to 85% of synchronous speed, the centrifugal switch opens and the motor will reach a speed as dictated by the load.

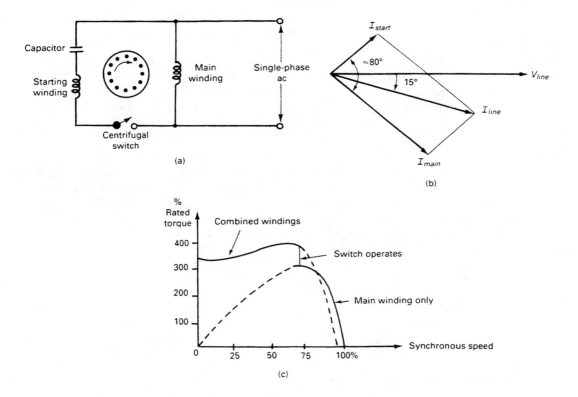

FIGURE 9.8. Capacitor-start induction motor: (a) circuit diagram; (b) phasor diagram; (c) torque-speed characteristic.

Because of their higher starting torques, capacitor-start motors are used in applications where not only higher starting torques are required, but also where reversible motors are needed. Applications of capacitor motors are in washing machines, belted fans and blowers, dryers, pumps, and compressors.

EXAMPLE 9.2

A capacitor is added to the starting winding of the motor in Example 9.1, with the result that its current now leads the voltage by 40°. The main winding remains as is.

a. With this added capacitor, determine at the instant of starting the line current and the power factor.

b. Compare the results with those calculated in Example 9.1.

Solution

a. From Example 9.1, $Z_{start} = \frac{120\angle 0°}{4\angle -15°} = 30\angle 15° = 28.98 + j7.76\,\Omega$. With the added capacitor, the starting winding impedance becomes

$$Z_{start} = 28.989 - j24.32 = 37.83\angle - 40° \Omega$$

Hence,

$$I_{start} = 3.17\angle 40° = 2.43 + j2.04 \text{ A}$$
$$I_{main} = 5.8\angle - 45° = 4.10 - j4.10 \text{ A}$$
$$I_{line} = I_{start} + I_{main} = 6.53 - j2.06 = 6.85\angle - 18° \text{ A}$$

power factor $= \cos(-18°) = 0.95$ lagging

Note that the value of the capacitor added is 82.7μ F.

b. The line current has been reduced from 9.48 A to 6.85 A and the power factor improved. The motor starting torque, being proportional to the sine of the angle between the winding currents, has also been increased and becomes maximum with minimum starting current. It can be shown that the starting torque of the motor with added capacitor, compared to that without, increases by a factor of

$$\frac{T_c}{T} = \frac{\sin[40° - (-45°)]}{\sin[45° - (15°)]} \times \frac{3.17}{4.0} = 1.58$$

where the subscript c indicates the developed torque with the added capacitor, and T that without.

Capacitor-Run Motors

In this motor, the capacitor is in series with the auxiliary winding is not switched out after starting. It remains in the circuit, and therefore the centrifugal switch is not needed. Since the capacitor is in the circuit continuously, it must be an ac paper oil type. The power factor is improved, as is the efficiency, but the capacitor value (on the order of 50 μF) is a compromise between best starting torque and running torque values, hence the starting torque is compromised.

Capacitor-Start Capacitor-Run Motors

The capacitor-start motor described has a high starting torque, but average running torque. For many applications this does not present a serious limitation.

In cases where high starting torques and high running torques are required, best results will be obtained if a large value of capacitance is used at startup which is then gradually decreased as the speed increases. In practice, two capacitors are used for starting and one is cut out of the circuit by a centrifugal switch once a certain speed is reached, usually at about 75% of full speed. The starting or intermittent capacitor is of fairly high capacity (usually on the order of 10 times the value of the running capacitor, for example, a 0.5 hp motor has $C_{st} = 300\mu F$, $C_{run} = 40\mu F$), which remains in the circuit. Figure 9.9 illustrates the connection diagrams for the capacitor-start capacitor-run motor, showing two methods generally encountered.

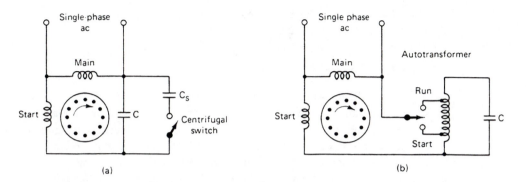

FIGURE 9.9. Capacitor-start capacitor-run motors: (a) two-value capacitor motor; (b) capacitor motor with autotransformer.

The first method, as shown in Figure 9.9a, uses an electrolytic capacitor in the starting circuit, which is not built to be left in the circuit continuously, since its leakage is too high. Being oil-filled, the second capacitor remains in the circuit; it has little leakage and therefore is suitable for continuous operation.

The second circuit (see Figure 9.9b) uses an autotransformer but only one oil-filled high-voltage capacitor. This method utilizes the transformer principle of reflected impedance from secondary to the primary. As we have discussed, this is proportional to the square of the secondary-to-primary turns ratio. For example, an autotransformer with 180 turns tapped at the 30-turn point would reflect an 8-μF running capacitor to the primary as

$$\left(\frac{180}{30}\right)^2 \times 8 \ \mu\text{F} = 288 \ \mu\text{F}$$

representing an increase of about 36 times. Thus a running oil-filled capacitor may be used for starting purposes as well, thereby eliminating one capacitor in lieu of the autotransformer, which is of comparable cost. Care must be exercised to ensure that the capacitor can withstand the stepped-up voltage, which is 180/30 = 6 times the rated voltage at startup. For instance, a 120-V motor would have a capacitor voltage at the instant of starting of 720 V. Typically, a 1000-V ac rating capacitor is required.

As is the case with the capacitor-start motor, the capacitor-start capacitor-run motor may be damaged for identical reasons if the centrifugal switch fails to operate properly. The primary advantage, then, of a two-value capacitor motor is its high starting torque, good running torque, and quiet operation. Reversing the line leads to one of the windings in the usual manner causes the motor to operate in the opposite direction. It is therefore classed as a reversible-type motor. In operations requiring frequent reversals it is preferred to use a single-value capacitor-run motor using no centrifugal switch.

Single-phase motors operate generally at relatively poor power factors, on the order of 0.5–0.6. Because of energy conservation awareness, considerable attention is being directed toward improving these low power factor devices. Presently, capacitor-start and capacitor-run single-phase motors have received the most attention, and as a consequence they now have power factors of up to 0.8.

Shaded-Pole Motors

Figure 9.10 illustrates the shaded-pole motor, consisting of a laminated stator core having salient poles and a concentrated main winding. The poles are divided in two parts, the smaller part of which is "shaded"; that is, it contains an auxiliary winding consisting of a single short-circuited turn of copper, called the *shading coil*. When the main winding is connected to an ac source, the magnetic field will sweep across the pole face from the unshaded to the shaded portion. This, in effect, is equivalent to an actual physical motion of the pole, the result is that the squirrel-cage rotor will rotate in the same direction.

To understand how this sweeping action occurs, let us consider the instant of time when the current flowing in the main winding is increasing most rapidly, as illustrated in Figure 9.11. The main flux ϕ will start to build up in phase with the current. However, the current induced in the shading coil produces an opposing flux, according to Lenz's law. The net result of this is that the flux in the shaded pole portion is less compared to that of the main portion of the pole. When the current in the main winding is at or near its maximum value, the flux does not change appreciably. With an almost constant flux, no voltage is induced in the shading coil and therefore it does not influence the main flux. The result is that the resultant magnetic flux shifts to the center of the pole. A short time later, when the current in the main winding is decreasing at its maximum rate, the flux in the unshaded portion of the pole decreases. However, because of currents induced in the shading coil, it tends to oppose this decrease in flux in the shaded portion of the pole. The result of this action translates into a movement of the magnetic flux axis toward the center of the shaded

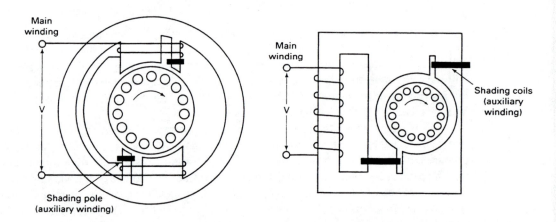

FIGURE 9.10.   Shaded-pole motor.

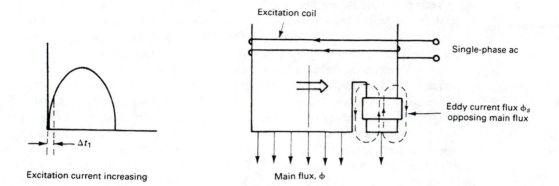

FIGURE 9.11. **The magnetic field in a shaded-pole motor, during the time the main-winding current is increasing most rapidly.**

portion of the pole. Hence, the flux ϕ_s continues to lag behind the flux ϕ during this part of the cycle.

It can similarly be reasoned that at any instant of the current cycle, the flux ϕ_s lags behind ϕ in time. The net effect of this time and space displacement is to produce a gliding flux across the pole face and consequently in the air gap, which is always directed toward the shaded portion of the pole. Therefore, the direction of rotation of a shaded-pole motor is always from the unshaded part toward the shaded part of the pole.

Simple motors of this type cannot be reversed, but must be assembled so that the rotor shaft extends from the correct end in order to drive the load in the proper direction. There are specially designed shaded-pole motors which are reversible.

Offsetting the simple construction and low cost of this motor are the low starting torque, low overload capacity, and low efficiencies (5 to 35%). These motors are built in sizes ranging from 1/250 hp up to about 1/20 hp. Typical applications of shaded-pole motors are where efficiencies are of "minor concern," such as in toys and fans. Since the applied voltage to the motor greatly affects its speed under load, as the slip increases with reduced voltage, advantage is taken of this fact, particularly when driving fans. Practically, this is generally done by providing line voltage taps to the excitation winding. With fewer turns on this winding, the volts/turn, as well as the current, is larger than with a full winding. Since the volts/turn ratio is proportional to the flux, it in turn increases and the motor runs at a greater speed, since a larger torque is developed.

9.2 Universal (Series) Motors

Motors that can be used on ac as well as dc sources are called *universal motors*. All these motors are of the dc series motor type. The direction of the developed torque is determined by both field polarity and the direction of the current through the armature. Since the same current flows through the field and the armature, it follows that on ac, reversals from positive to negative, or vice versa, will simultaneously affect both the field direction and the current direction through the armature. This means that the direction of the developed torque will remain the same, and rotation will continue in the same direction. Although the torque is in the same direction, it pulsates in magnitude at twice the line frequency. The torque-speed characteristic at ac and dc operation differ somewhat, as illustrated in Figure 9.12. This is attributable to the relatively large voltage drop across the series field due to its high reactance on ac operation, which reduces the output power. The reduction in output power can be compensated for by providing more armature winding turns. However, this in turn increases commutation problems. Thus a universal motor has to be designed carefully to minimize these drawbacks. Factors such as increased brush area, brush material, and pressure are fairly critical. Another difference between the universal motor and an ordinary dc series motor is in the parts of the field structure. On ac operation the field flux induces eddy currents in the solid parts of the field structure, such as the yoke and cores, generating excessive heating and thereby lowering the efficiency. This is overcome by laminating these parts and operating the machine generally at a lower flux density and using very short air gaps. Because of the relatively small number of field ampere-turns and the low flux density, a short pole with a large cross section results.

Rotor speeds of universal motors are generally in the range 5000 to 20,000 r/min, the higher speeds occurring at no load. These high no-load speeds are not destructive since the attached gear box usually prevents the motor from overspeeding and thus reaching these speeds, or because the load is attached directly. Universal motors for small power applications such as portable hand drills and food mixers operate at high speeds but are geared down to their loads so that they operate at conveniently low values. The motors in these applications operate at high speeds because, other things being equal, at higher

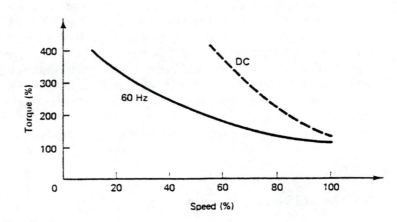

FIGURE 9.12. Typical torque-speed characteristic of a universal motor.

speeds better cooling is obtained. When properly designed, the gear losses are kept to a minimum. In other applications, the actual motor speed is the load speed; an example of this is the vacuum cleaner. Since all commutator-type motors, especially those operating at high speeds, interfere with radio reception, universal motors are usually provided with a small capacitor connected directly across the line terminals inside the motor.

In addition to the applications mentioned above, there are numerous applications where universal motors are used, such as portable drills, hair dryers, hedge trimmers, polishers, routers, and sewing machines, just to name a few. Applications must take into consideration its large speed regulation as well as the large horsepower developed for its size.

Speed control of universal motors is best obtained by solid state devices. These electronic devices consist of thyristors or triacs in wave-chopping circuits to provide voltage variation to the series motor. As a consequence, speeds from full load to almost zero can be achieved with the motor developing full load-torque.

9.3   *Permanent-Magnet DC Motors*

Permanent-magnet dc motors have their pole fields supplied by permanent magnets (see Figure 9.13). The semicircular permanent magnets are assembled on the stator frame to provide one or more field pole pairs. Most permanent magnets are composed of either ceramic, Alnico, or rare earth elements. Without a field-wound structure, permanent-magnet dc motors become more compact in size and tend to have higher efficiency. Elimination of field-coil power loss results in cooler operation with totally enclosed constructions. Alnico-type magnets are least affected by higher temperatures. Disadvantages are the absence of field control and special speed-torque characteristics. If permanent-magnet motors are not overloaded, demagnetization of the air-gap flux will not occur and the torque-speed characteristic curves remain essentially linear. All torque-speed control is achieved by adjustment of the armature voltage. Below rated load, the operating characteristics are linear and directly proportional to the armature voltage. The operating characteristics are illustrated in Figure 9.13b. In summary, permanent-magnet motors provide comparatively simple and reliable dc drives where high efficiency, high stall torque, and linear speed-torque operating characteristics are desirable. Other advantages are the reduced physical size and availability of totally enclosed frames.

Analysis

The permanent-magnet dc motor of Figure 9.13 can be represented by an equivalent circuit shown in Figure 9.14a. The counter EMF and armature current are expressed by

$$E_c = k_c\,\omega \tag{9.1}$$

$$I_A = \frac{V_L - E_c}{R_A} \tag{9.2}$$

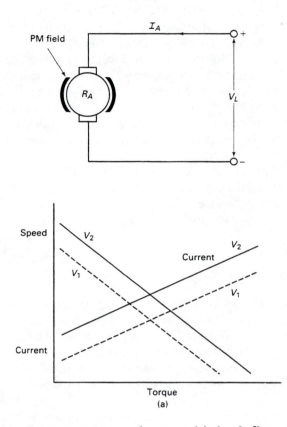

**FIGURE 9.13.   Permanent-magnet dc motor: (a) circuit diagram; (b) speed
and current versus torque curves.**

where k_c is the counter EMF constant, V/rad/s. The motor torque is given by,

$$T = k_T I_A$$

$$= \frac{k_T(V_L - E_c)}{R_A} = \frac{k_T(V_L - k_c\omega)}{R_A} \tag{9.3}$$

where k_T is the torque constant, N·m/A. The curve of speed versus torque will be a straight
line as indicated by Eq. 9.3. Transposing this equation yields

$$T R_A = k_T V_L - k_T k_c \omega \tag{9.4}$$

and

$$\omega = \frac{k_T V_L - T R_A}{k_T k_c} = \frac{V_L}{k_c} - \frac{R_A}{k_T k_c} T \tag{9.5}$$

This linear relationship of speed versus torque is illustrated by Figure 9.14b.

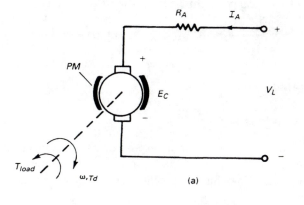

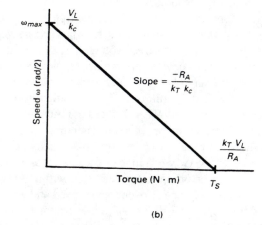

FIGURE 9.14. Analysis of permanent-magnet dc motor: (a) equivalent circuit; (b) torque-speed characteristic.

At standstill $\omega = 0$; then from Eq. 9.5,

$$T = T_{start} = \frac{k_T V_L}{R_A} \tag{9.6}$$

The maximum speed occurs when T = 0:

$$\omega_{max} = \frac{V_L}{k_c} \tag{9.7}$$

The developed power that will be converted to mechanical power is

$$P_e = E_c I_A = k_c \omega I_A \tag{9.8}$$

and the mechanical power to the load when $P_{rot} = 0$, is

$$P_d = \omega T = \omega k_T I_A \tag{9.9}$$

9.4 Servomotors

DC Servomotors

Dc servomotors are used primarily in position and velocity control applications. These vary from 1 A motors developing 0.0023 N·m used in photography cameras to 2 hp motors used in antenna drives, moving equipment in automatic warehouses and position actuator arms in robots.

In velocity control devices, for a fixed voltage torque is relatively constant with speed, similar to most large dc motors. The major difference for a servomotor is that it will usually be better built and exhibit very little torque ripple as it has a larger number of commutator segments. Normally it is controlled from a variable voltage source in contrast to most machines which operate at a fixed voltage.

Position control devices are mainly employed in high torque, low speed situations. The torque-speed curve is much steeper than a conventional dc motor because it produces very high torque at standstill, but relatively little torque at high speeds. These motors are also called *torque motors*. We will focus our attention on a typical application for a small permanent-magnet dc motor. A set of operating characteristics for a 2 hp direct-drive dc motor is shown in Figure 9.15. Motors specifically designed for direct drive applications can be used over a very wide range of speed and torque without requiring the use of gears. These motors can be brought to a dead stop within 3 revolutions from 1000 r/min. Since the dc servomotor will be run over a wide speed and voltage range there is no fixed voltage curve, and it then becomes necessary to determine the power dissipation in the motor (see Figure 9.15). Note that there are three different regions in which the motor can operate:

1. Continuous duty zone.
2. Safe acceleration zone.
3. Intermittent zone.

The specific conditions under which the motor can operate in zones 2 and 3 are detailed in manufacturers' data sheets. Figure 9.16 shows the torque-speed characteristic of a dc servomotor used in a position-control device. This would be an application where an output has to move at very low speed and is essentially stationary. One example is a crystal pulling machine employed in the making of pure silicon for semiconductors. Motor speed must be very accurately controlled and the speeds are very low.

Note the curves in Figure 9.16. The particular data were obtained from a permanent-magnet motor. The torque-speed curves are extremely linear since the magnets—in this situation ceramic magnets—are virtually unaffected by armature reaction. Servomotors are usually higher resistance machines than conventional motors, since they may require high

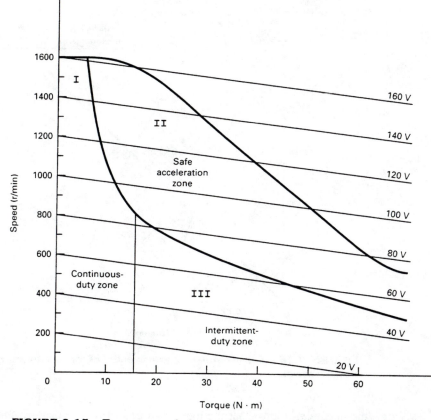

FIGURE 9.15.   Torque-speed characteristic of a 2 hp direct-drive dc servomotor.

currents at relatively high voltages in order to produce large torques. For this reason, and also because only a small amount of cooling may be obtained from air motion through the machine, the frame must be comparatively large. This allows heat to be dissipated over a larger surface area.

EXAMPLE 9.3

To show the use of Figure 9.16, consider a given process that has the following torque-speed characteristic:

$$T = k\omega$$

where ω is the speed (rad/s), and the machine constant $k = 96.1 \times 10^{-6}$ N·m $= 10^{-4}$ N·m (r/min). Determine

a. the motor voltage when the load is rotating at 2000 r/min,

b. the starting torque available at the voltage determined in part (a), and

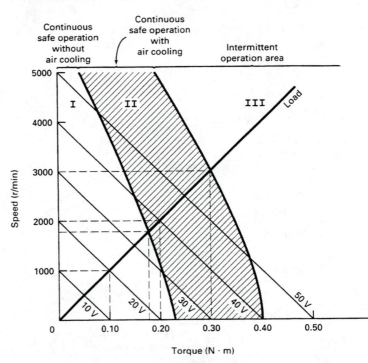

FIGURE 9.16. **Torque-speed characteristic of dc servomotor for positional control.**

c. the maximum torque that the motor can deliver to the load, assuming it can be air cooled.

Solution   Figure 9.16 shows the load curve superimposed on the motor characteristic curves. From this we can see that:

a. At 2000 r/min, the load torque is 200×10^{-3} N·m or 0.2 N·m. For a speed of 2000 r/min and a torque of 0.2 N·m, the voltage is approximately 40 V.

b. From the curve, the starting torque is 0.4 N·m.

c. Maximum torque with air cooling occurs at the zone II–zone III boundary and is about $300 \times 10^{-3} = 0.3$ N·m. The speed is 3000 r/min. At this operating, the power output is

$$P = \frac{3000 \times 2\pi}{60} \times 0.3 = 94 \text{ W}$$

Note: The operating point in (a) is in zone II. This is acceptable for intermittent operation or when the motor is air cooled. For continuous operation without air cooling (to avoid overheating), the maximum speed obtainable occurs at the point where the load line meets the zone I—zone II boundary. This occurs at about 1900 r/min, 36 V.

AC Servomotors

The ac servomotor as used in position-control systems is basically a two-phase induction motor. The stator has two field windings wound on pole pieces and at right angles to each other. The rotor is an ordinary squirrel-cage type. The arrangement of the stator construction is shown in Figure 9.17.

One of the windings, called the *reference winding*, is connected to the ac supply, the other, the *control winding*, is connected to some other voltage supply, being of the same frequency as the reference winding. If a voltage phase difference exists between the two voltages applied to the stator windings, a revolving field will be created and the rotor will rotate in the fashion discussed. The sense of direction will depend on the relative phase difference between the two voltages. Thus the design enables the motor speed or stalled torque to be controlled by regulating one phase, keeping the other fully energized.

The motor characteristics differ considerably from those of ordinary motors, since the torque-speed curve is approximately linear. By reducing the voltage to the control winding, a series of speed curves can be obtained as shown in Figure 9.18. To assure high torques and acceleration, the rotor length is relatively large compared to its diameter. Also, the rotor resistance is high to prevent the motor from running single-phase after the control voltage has been reduced to zero.

9.5   Moving Coil Armature Motors

The moving-coil armature type motors are constructed of permanent-magnet field poles on the stator with either a shell or hollow-cup armature winding, or a printed-circuit disk

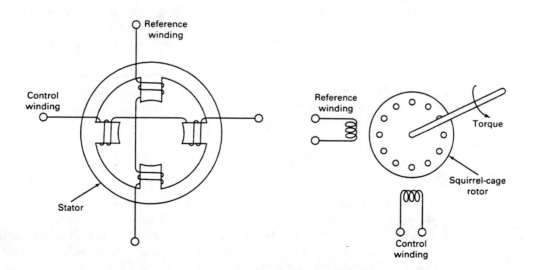

FIGURE 9.17.   Two-phase servomotor.

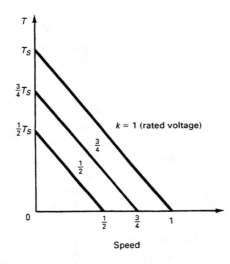

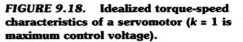

FIGURE 9.18. **Idealized torque-speed characteristics of a servomotor ($k = 1$ is maximum control voltage).**

armature winding. Commutator segments are connected to the armature windings which are assembled on the rotor. Within the armature structure there is no iron or steel. Thus, the armature is of extremely light weight resulting in very low inertia. Also, the armature windings possess low inductance and correspondingly low electrical time constants. As a consequence, the moving-coil armature type motors are used in applications requiring rapid response due to low inertia and high acceleration. The high acceleration is produced by the high starting torque which in turn is the result of a high value of armature current. Because of the armature design, this high value of armature current must be limited to short durations in order to prevent overheating. Shell-type armature windings composed of insulated copper or aluminum coils are assembled and bonded using resin to form a hollow cup. The hollow cup winding is connected to the commutator assembly and mounted on the rotor shaft as illustrated in Figure 9.19a.

Printed-circuit armature windings composed of laminated or etched tracings of copper form the conductors on a fiber glass disk which are connected to the commutator assembly. The printed-circuit winding and commutator segments are mounted on the shaft and assembled as shown in Figure 9.19b.

9.6 Brushless DC Motors

Where there is a need for a dc motor in which the hazards produced by sparking or arcing at the brushes and commutator segments must be eliminated, the brushless dc motor is finding increased application. Also, applications requiring constant-speed drives (such as hard-disk drives), often have brushless dc motors installed. Essentially, the brushless dc motor consists of a permanent magnet field rotor and stator armature windings which are commutated by means of electronic switching. Effective rotation of the armature field is

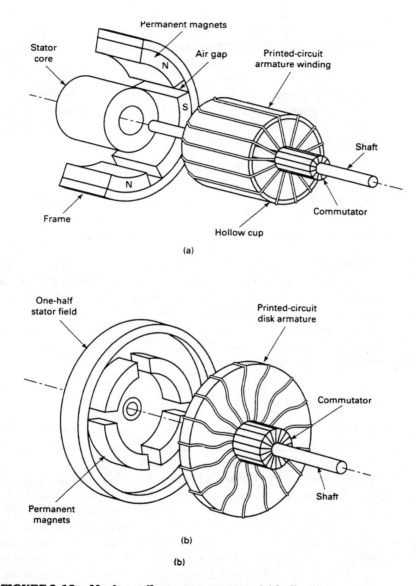

FIGURE 9.19.   **Moving-coil armature motors: (a) hollow-cup armature motor; (b) printed-circuit disk armature motor.**

achieved by changing the current directions in the armature windings by means of switching power transistors. To provide synchronism of the permanent magnet field of the rotor with the rotating armature fields of the stator, shaft position sensors are used to initiate the proper switching times. The electronic circuitry for brushless dc motors may be quite complicated;

however, the general principles are illustrated in Figure 9.20. As shown, a three-phase brushless dc motor control system consists of three groups of armature stator coils which are excited by electronic driver circuitry controlled by the position sensors. The structure of the brushless dc motor is thus similar to that of a synchronous motor. The number of field poles is determined by the permanent-magnet field structure. Electronic control of both magnitude and switching rate of armature current in the stator windings will determine the inherent torque-speed characteristics of the particular brushless dc motor control system. Brushless dc motors with their constant speed versus torque characteristics are becoming competitive with conventional dc drives.

9.7 Tachometers

DC Tachometers

Dc tachometers are precision small generators designed to develop a voltage output directly proportional to angular velocity. Many tachometers are applied to feedback control systems in which the tachometer supplies an accurate voltage sensing signal to the closed-loop system of speed regulation. Generally, tachometers are permanent-magnet generators with suitable numbers of commutator segments and armature coils. The armature design is such that output voltage ripple over the desired speed range is negligible. Tachometers are designed for continuous operation.

AC Tachometers

To measure the angular velocity of a shaft, a small two-phase generator with a permanent magnet may be used. Its output can then be measured with a voltmeter which is calibrated in terms of speed. Unfortunately, in such a device the measured output voltage does not only vary with speed, but its frequency is speed-dependent as well. In many applications, such as servo systems using synchros, the input and output variables are at a fixed frequency and only the voltages change. If the output from a tachometer is to be added directly as an ac signal, its output frequency must also be fixed. Thus it is clear that a conventional generator (alternator) cannot be used in such an application. To overcome this frequency dependency on speed, a machine known as a *drag-cup generator* is often employed, illustrated schematically in Figure 9.21. Basically, the tachometer generator consists of an input winding energized from an ac source. This winding is so arranged as not to induce a voltage in a separate output winding. A copper or aluminum sleeve, the drag-cup rotor, runs in the air gap and is supported by a shaft. By transformer action eddy currents are induced in the rotor, which in turn set up an armature reaction flux at right angles to the field set up by the reference winding (see Figure 9.22).

The resultant field, being shifted in space in the direction of rotation, will induce a voltage in the output winding. If the field due to the reference winding is constant, the induced armature currents, and hence its created armature flux, will depend directly on the

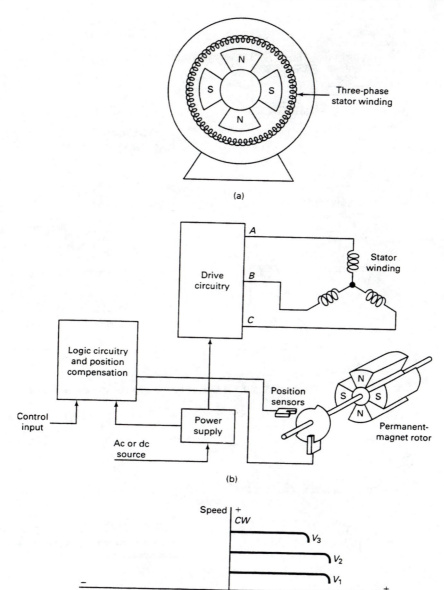

(a)

(b)

(c)

FIGURE 9.20. Brushless dc motor: (a) construction; (b) control and drive circuitry; (c) torque-speed characteristic.

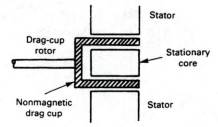

FIGURE 9.21. Drag-cup tachometer; rotor cross section.

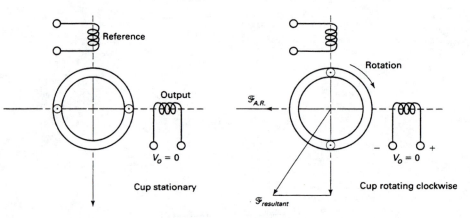

FIGURE 9.22. Field relations in drag-cup tachometer.

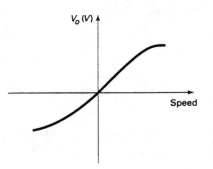

FIGURE 9.23. Drag-cup tachometer output voltage V_o versus speed characteristic.

rotational speed of the motor. Figure 9.23 shows the voltage-speed characteristic for the tachometer.

9.8 Stepper Motors

Stepper motors are special motors that have the ability to translate an electrical input signal (usually a pulsed square-wave) into discrete angular increments in shaft position. It has the ability to rotate in either direction, as well as stop and start at various shaft positions. The motor is designed to rotate a specific number of degrees for each electrical pulse it receives from its drive unit. For a given number of pulses the shaft will have turned through a known angle. The angle through which the shaft moves for each pulse is called the *step angle*, expressed in degrees. Depending on its design, a stepper motor can advance by 45°, or as little as a fraction of a degree per pulse. The smaller the step angle, the greater the number of steps per revolution, and the higher the resolution of position.

Stepper motors are manufactured in subfractional to integral horsepower ratings, and their applications are numerous. The behavior of a stepper motor depends to a large extent on the power supply that drives it. The power supply provides the pulses which, in turn, are usually initiated by a microprocessor or computer.

Stepper motors are widely used in numerical control for machine tools, in driving paper feed mechanisms in line printers, X-Y plotters, such as positioning a worktable in two positions for automatic drilling, in robots applications where the torque requirements are relatively small, magnetic head positioning in floppy disk drives where they provide precise positioning of the magnetic head on the disks, and so on. The stepper motor is ideally suited for such applications because it is a device which converts input information in digital form to an output that is mechanical, namely, rotational motion. It provides a natural interface with digital computers. The stepper motor allows control of position, velocity, distance, and direction. Because each step moves the shaft to a known location, the only shaft position error, regardless of distance travelled, is generally negligible, since the positional error is noncumulative. The number of steps in each revolution of the shaft varies depending on the type of motor, which in turn is dictated by its intended use. Typical stepping angles range from 1.5 to 30°. Stepper motors have inherent low velocity without gear reduction. A typical unit driven at 500 pulses/s turns at 150 r/min.

There are three types of stepper motors:

- variable-reluctance stepper motors
- permanent magnet stepper motors
- hybrid stepper motors

We will discuss the variable-reluctance stepper motor and the hybrid stepper motor since they are by far the most common. They have relatively high torques per unit volume and have smaller stepping angles than a comparable permanent-magnet stepper motor.

Variable-Reluctance Stepper Motor

Figure 9.24 shows a cross-section of a variable-reluctance stepper motor. There are three phases, arranged on six stator poles; there are four rotor poles. The diagram shows three successive rotor positions in response to energizing phases A, B, and C in sequence; the step angle is 30°. The rotor follows the air-gap field around by virtue of the reluctance torque. This reluctance torque is generated because of the tendency of the ferromagnetic rotor to align itself along the direction of the resultant field.

We can reverse the direction of rotation by energizing the phases in reverse sequence, namely, A, C, B, A .... In order to fix the final position of the rotor, the last phase that was

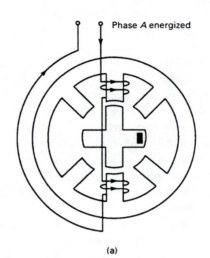

Phase *A* energized

(a)

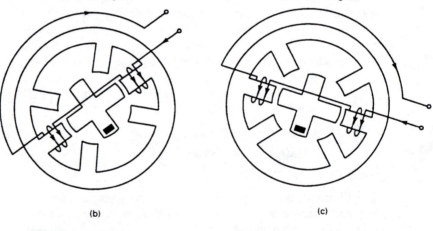

Phase *B* energized

Phase *C* energized

(b)   (c)

FIGURE 9.24.   Shaft motion of a variable-reluctance stepper motor in response to switching excitation from: (a) Phase *A*; (b) to Phase *B*; (c) to Phase *C*.

energized in the sequence must remain energized. This holds the rotor in its last position and prevents it from moving under the influence of external forces. In this stationary state, the motor will remain "locked" provided the external torque does not exceed the so called *holding torque* of the motor.

Smaller step angles can be achieved in a number of ways; using more stator and rotor poles; employing multistack constructions; or energizing the stator poles in the so-called microstepping mode, whereby current is balanced in the stator phases of the motor. For example, energizing Phase A and B simultaneously with identical currents results in a step of 15° for the motor in Figure 9.24a. However, this mode of operation requires more complex integrated circuitry.

Multistack Stepper Motor

Multistack variable-reluctance stepper motors are widely used to provide smaller step angles. As Figure 9.25a shows, there is a separate stator and rotor combination for each phase. The rotor has small salient poles or teeth, resembling a gear-like arrangement. The stator has four poles with similar teeth formed in the pole faces, as schematically indicated in Figure 9.25b.

When Phase A is energized, all the rotor teeth in that stack are pulled into alignment with the stator teeth. Both rotors are mounted on the common shaft, and the stators in successive stacks are displaced by an angle which equals the step angle. Switching the current from Phase A to Phase B will cause the next set of rotor teeth to align with its corresponding stator teeth. In doing so, the shaft rotates through one step in changing excitation from Stack A to Stack B. Another step motion in the clockwise direction can be obtained if excitation is changed from Stack B to Stack C. Another change of excitation from Stack C to Stack A will align the stator and rotor teeth in Stack A once more. However, during this process $(A \rightarrow B \rightarrow C \rightarrow A)$ the rotor has moved one rotor tooth pitch, that is, the angle between rotor teeth. Letting x be the number of rotor teeth and N the number of stacks or phases, then

$$\text{Tooth pitch} \quad \tau_p = \frac{360°}{x} \tag{9.10}$$

and

$$\text{Step size} \quad \Delta\theta = \frac{360°}{xN} \tag{9.11}$$

Typical step sizes for the multistack variable-reluctance stepping motor are in the range 2° to 15° degrees.

The advantage of the variable-reluctance stepper motor is that the torque developed by the motor depends on the magnitude of the current in a phase and not its direction. The magnetic alignment force does not depend on the field direction. This simplifies the drive circuitry considerably, since the current needs only to be switched on and off and does not have to be reversed.

Hybrid Stepper Motors

To augment the variable-reluctance torque, a permanent magnet is used, resulting in a construction schematically indicated in Figure 9.26. This motor is classified as a *hybrid stepper motor*; it combines the variable-reluctance and permanent-magnet stepper motor characteristics. Smaller stepping angles can be obtained with this motor compared to that of the variable-reluctance stepper motor, but it is more expensive.

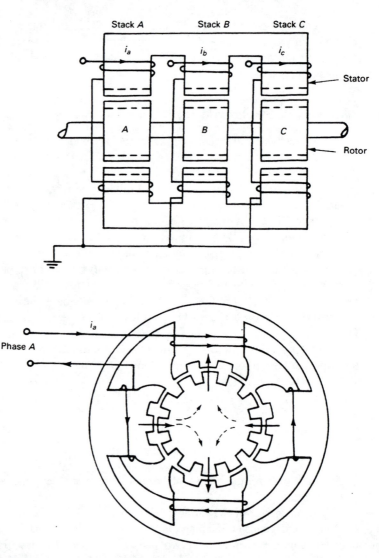

FIGURE 9.25. Three-stack variable-reluctance stepper motor; (a) cross section; (b) teeth position for Phase A excited.

The stator construction is similar to that of the variable-reluctance motor, but the rotor construction is different. An axial view of this motor is illustrated in Figure 9.26a. As shown, the rotor is made up of an axial permanent magnet sandwiched between two ferromagnetic gear-like rotor sections, and there is an angular displacement between the teeth of the two rotor parts (in this arrangment it is 20°). These outside rotor sections consist of equal number of teeth. The stator winding of both phases bridge the stator poles of both

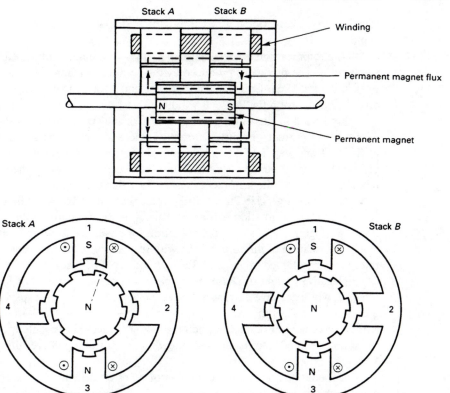

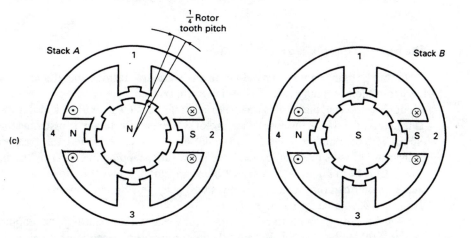

FIGURE 9.26. Hybrid stepper motor: (a) longitudinal cross section; (b) cross section showing rotor poles: Phase A energized; (c) Phase B energized.

stacks. The cross-sectional views through the Stacks A and B are shown in Figures 9.26b and c.

The rotor teeth in Stack A act like north poles, and the rotor teeth in Stack B act like south poles. The flux paths produced by the permanent magnet are indicated by the dotted lines in Figure 9.26a. Coils on stator poles 1 and 3 form the winding of Phase A, and coils on pole 2 and 4 form the Phase B winding.

Figure 9.26b shows the rotor position for the condition when Phase A is energized with positive current. Positive current of Phase A causes the stator pole flux in gaps 1 of Stack A and gap 3 of Stack B to aid the permanent-magnet flux, but it opposes the permanent-magnet flux in gaps 1 of Stack B and gap 3 of Stack A; the rotor teeth are pulled into alignment with the stator teeth where the field is strongest.

If we now energize Phase B, the rotor will rotate 10° and line up with stator poles B. The direction of rotation will again depend on the direction of the current flow through Phase B. Figure 9.26c shows the condition for positive current in Phase B, resulting in a clockwise rotation of the rotor of 10° with respect to its previous position. As shown, the rotor teeth are aligned in gap 2 of Stack A and gap 4 of Stack B.

Negative current in Phase A will cause alignment in gaps 3 of Stack A and gap 1 of Stack B, while negative current in Phase B causes alignment in gaps 4 of Stack A and gap 2 of Stack B. Thus the current sequence $+A \rightarrow +B \rightarrow -A \rightarrow -B \rightarrow$ will result in a rotor movement of one tooth pitch (i.e., 40°). Practical two-pole motors usually have 8 stator poles and 50 rotor teeth, with 5 teeth on each stator pole, resulting in a step angle of 1.8°.

It should be noted that the number of poles on the stator of a stepper motor is never equal to the number of poles on the rotor. This feature is totally different from any other type of motor. It is the difference in the number of poles that enables the motors to step as they do.

Stepper Motor Characteristics

When the stepping motor steps along in the start-stop fashion there is an upper limit to the permissible stepping rate. If the pulse rate of the driver sending current to the motor windings is too fast, the rotor is unable to accurately follow the pulses and steps will be lost. This defeats the whole purpose of the motor which is to correlate its instantaneous position (steps) with the number of net (+ and −) pulses. As can be expected, the greater the inertia of the rotor plus its mechanical load, the slower the stepping rate to bring the motor up to speed. Curve I in Figure 9.27, the *pull-in torque* curve, shows the allowable load which the stepper motor can run without missing steps. Any pulsing rate (speed) in the shaded area would be acceptable for the given load.

If the motor is made to run at a uniform speed without starting and stopping at every step, it is said to be *slewing*. Because the motor then runs essentially at uniform speed, the effect of inertia is absent. Consequently, the motor can carry a greater load torque when it is slewing. However, should the load torque increase, there is a point at which the motor will fall out of step and the position (steps) of the rotor will no longer correspond to the net number of pulses delivered to its windings. The *pull-out torque* curve in Figure 9.27 shows

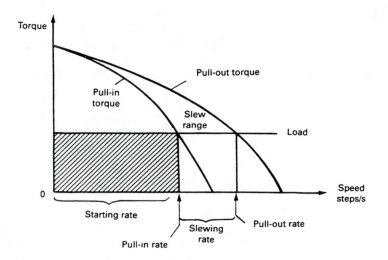

FIGURE 9.27. Typical torque-speed curve of a stepper motor.

the allowable load which the motor can run at its maximum stepping rate, after it has been slewed up to speed.

When a motor is carrying a load, it cannot suddenly go from zero to some large stepping rate. Likewise, a motor that is slewing at a high speed cannot be brought to a dead stop in one step. To bring the motor up to speed it must be accelerated gradually; conversely, when running at full speed it must be decelerated gradually—within the constraint that the rotor position corresponds to the number of pulses. This process of acceleration and deceleration is called *ramping*. Usually the ramping phase is completed in a fraction of a second and is generated by the motor driver.

9.9 Electromagnetic Clutches

In the operation of controlled devices it is often desirable to be able to quickly engage or disengage a motor from the device it is driving. This is the function of a clutch. There are numerous types of clutches; only two of the most commonly used clutches are discussed here.

Friction Disk Clutch

The basic principle of this device is illustrated in Figure 9.28. The input shaft and the clutch disk fastened to it are rotated by the motor. The second disk is fastened to the output shaft, which in turn is connected to the actuated device. As shown in the figure, the two disks are separated and the output shaft does not rotate. As the two disks are pressed together,

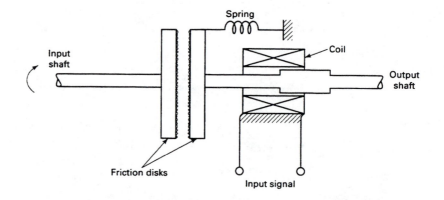

FIGURE 9.28. **Operating principle of a friction clutch which engages upon energizing coil.**

the output shaft is made to rotate because of the friction between the surfaces of the disks. When separated, the output shaft is uncoupled from the input shaft and rotation ceases.

There are various ways possible to bring the two clutch disks together, the usual one is electromagnetically. As shown in Figure 9.28, the disks are kept separated by a spring. Upon energizing the coil, the magnetic field set up by the coil current will actuate one of the clutch disks against the spring action, causing it to engage with the other. Interrupting the coil current causes the spring force to disengage the clutch disks.

The reverse arrangement, where the disks are normally engaged, is shown in Figure 9.29. Here, upon energizing the coil the magnetic field will create a force to pull the disks apart, thereby uncoupling the shaft against the spring action. When the input signal has been removed, the spring force will keep the disks normally engaged.

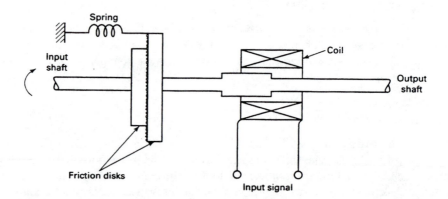

FIGURE 9.29. **Friction disk clutch which disengages upon energizing coil.**

It should be observed in Figures 9.28 and 9.29 that the input and output shafts refer to the clutch arrangement. In actual practice, the motor and controlled device shafts are attached to the input and output shafts of the clutch, respectively, via a shaft coupler.

Eddy Current Clutch

The electromagnetic clutch discussed has two stable positions, that is, it is an on-off device; input-coupled or not coupled. When coupled, the output shaft rotates at the same speed as the input shaft.

Many industrial operations, however, require that the speed of the actuated device be controlled, a function the friction clutch cannot provide. There are several types of clutches available that can perform this function, one of which is the eddy current clutch. The basic principle on which this device is based is as follows. Figure 9.30 shows a simple electromagnet arrangement consisting of two U-shaped iron bars mounted on a shaft. Upon energizing the coil with a dc current through a slip ring arrangement, magnetic poles are created, setting up a flux as shown. This arrangement is now inserted in a soft-iron sleeve or drum, preserving a small air gap, and the magnet arrangement rotated. The moving poles, and hence the rotating magnetic flux linking the drum, set up eddy currents in the drum. These eddy currents create a secondary field which will interact with the rotating magnet field. The resulting force is in such a direction as to cause the drum to follow the magnet field (see Figure 9.31). It can now be imagined that the magnet arrangement is coupled to the input shaft and the drum to the output shaft, with no mechanical connection between the two shafts.

The torque created by the force between the magnet field and eddy current field can be controlled by varying the dc excitation current of the coil, thereby controlling the rotational speed of the drum. Because eddy currents are only created when there is relative motion between the poles and drum, it is inherent in the device operation that the rotational speed

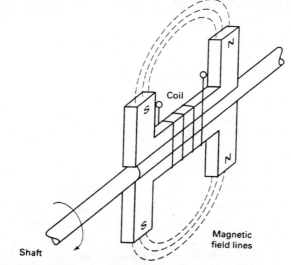

Coil

N

S

N

S

Magnetic field lines

Shaft

FIGURE 9.30.   Eddy current clutch. Magnetic sleeve, enclosing this assembly, has been omitted for clarity.

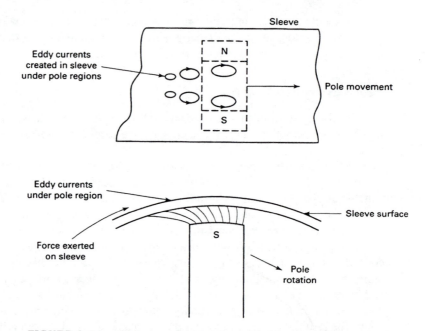

FIGURE 9.31. **Eddy current clutch action illustrated in principle.**

of the output shaft cannot be synchronized with that of the input shaft. In other words, eddy current clutches require slip to function, which normally do not exceed values of 5% at rated torque. It is interesting to note that the torque-speed characteristic of an eddy current clutch is similar to that of an induction motor. It develops sufficient peak torque to prevent stalling in case of momentary overload, and its characteristic curve can be influenced by selection of drum material. This is analogous to adding resistance in the rotor circuit of wound-rotor induction motors.

PROBLEMS

9.1 A small motor is rated 10 W at 3050 r/min; what would be its developed torque in newton-meters? In pound-feet? In ounce-inch?

9.2 A single-phase 115-V 1/3-hp 1720-r/min motor draws a current of 5.2 A when fully loaded. Calculate the efficiency at full load and PF = 0.72.

9.3 A single-phase induction motor has a main winding impedance of $6\angle 45°$ Ω and an auxiliary winding impedance of $10\angle 15°$ Ω.

    a. What capacitor value is needed in series with the starting winding to produce a 90° phase shift between winding currents at startup?

    b. With this added capacitor, what are the winding currents and line current at starting? The line voltage is 120 V.

9.4 A four-pole shaded-pole motor has 165 turns per coil wound for 115 V. All four coils are connected in series. The motor is to be operated from a 230-V source, necessitating reconnection of the windings. What changes would be made if the volts per turn are to remain the same?

9.5 An autotransformer and a 4-μF capacitor combination are used in a two-value capacitor motor. If the transformer ratio at startup and run is 6 and 1.5, respectively, determine the effective capacitor values when viewed from the primary side of the transformer and the minimum ac voltage rating needed. The line voltage is 120 V.

9.6 A small dc permanent-magnet motor draws 1 A at 12 V, and runs at 600 r/min. The armature resistance is 1 Ω. What value of resistance should be added to the armature circuit to drop the speed to 450 r/min, assuming the armature current remains constant.

9.7 The torque-speed characteristic of a dc servomotor used as a position-control device is as given in Figure 9.32. The particular data shown were obtained from a permanent-magnet motor. The characteristics are linear since the magnets in this case are ceramics and are virtually unaffected by armature reaction. It is used to drive a load having a torque-speed characteristic as $T = k_1 n$, where n is the speed in r/min, T is the torque in N·m, and $k_1 = 64 \times 10^{-6}$ N·m/rad/s $= 0.67 \times 10^{-4}$ N·m/(r/min). Determine

   a. the motor voltage when the load is rotating at 2000 r/min,
   b. T_{start} at the voltage determined in part (a),
   c. n_{max} if no external cooling is used,
   d. the power to the load in part (a), and
   e. T_{max} the motor can deliver if air cooled.

9.8 The torque-speed characteristic curve of a dc servomotor is given in Figure 9.32. It is to be used to drive a load having a torque-speed relationship given by

$$T = 4 \times 10^{-6} \omega^2$$

where T is the torque in N·m and ω is the speed in rad/s. Determine

   a. the maximum voltage that may be applied to the motor if it is not fan-cooled,
   b. the speed at which the motor will operate if the applied voltage is 25 V,
   c. the power output of the motor if it is driving the load at 250 rad/s, and
   d. the range of the voltage applied if the motor is to be fan-cooled and operate within safe limits.

9.9 A 115-V universal motor is loaded and takes 6.8 A at a power factor of 0.866 lagging. The output torque is measured to be 0.5 N·m at a speed of 4000 r/min. A resistor, used to control the speed, is connected in series with the motor, resulting in a voltage drop across the motor of 108 V and a line current of 8.5 A at a power factor of 0.80 lagging. The produced torque is maintained but at a lower speed. The motor impedance is 0.6 Ω. Determine

   a. the speed the motor is operating at with the series resistor added, and
   b. the efficiency the motor is operating at with and without this added resistor.

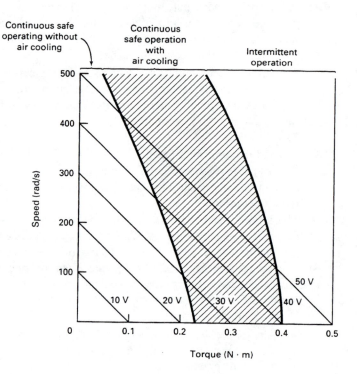

FIGURE 9.32.   Torque-speed characteristic of a dc servomotor.

9.10   A universal motor has a resistance of 15 Ω and an inductance of 0.21 H. When connected to a 120-V dc supply and loaded, it takes 0.8 A and runs at 1800 r/min. When connected to a 120-V 60-Hz supply and loaded, it takes the same current but at a lagging power factor of 0.866. Determine

   a.   the speed at which the motor will run, and
   b.   the shaft torque in N·m if the rotational losses amount to 10 W.

Assume the impedance voltage drop is in phase with the applied voltage.

9.11   A series motor with an armature circuit resistance of 0.4 Ω takes 50 A at 250 V while hoisting a given weight at 300 ft/min. Assuming linearity, specify the resistance to be placed in series with the motor to slow the hoisting speed to 200 ft/min.

9.12   A dc servomotor has a voltage constant $k_g = 0.824$ V·s/rad and a torque constant $k_m = 7.29$ lb.in/A. The armature resistance is 0.41 Ω. Determine the steady-state speed of the motor at no load and full load (120 lb.in) conditions. The line voltage is 36V.

9.13 A 4-ft robot arm is lifting a 25-lb weight at 60°/s. Assume the arm has zero mass. Determine the hp rating of the servomotor.

9.14 A permanent magnet dc motor is used to drive a robot joint that requires a torque of 50 in.lbs. If the motor constant $k_m = 2.5$ in.lbs/A, how much current is required from the drive amplifier at peak torque?

9.15 A single-stack, four-phase stepper motor is required to produce an 18° step motion. Determine the number of rotor poles and the sequence of excitation of the stator phases. Draw a cross-sectional view of the motor.

Synchronous Motors

10.1 Introduction

In earlier discussions we have seen that generators and motors are electrically and magnetically identical. A motor can be operated as a generator, and vice versa. Only minor changes in nameplate ratings are involved in doing so.

In the same way we expect that a synchronous generator can be operated as a synchronous motor. If the prime mover is uncoupled from the synchronous generator, a three-phase supply of constant frequency connected to its armature winding while the field is provided with dc excitation, torque will be developed at synchronous speed. Putting a mechanical load on the shaft will not cause the motor speed to change. The speed regulation is thus zero, and we have a constant speed machine. There are many industrial applications requiring constant speed, such as compressors, pumps, and mill machinery.

Another important advantage of synchronous motors is their adjustable power factor. They can provide correction for low power-factor caused by other types of motors. The synchronous motor's power factor is changed by merely changing the dc field excitation current. When the motor power factor is unity, the dc excitation is said to be normal. Overexcitation causes the motor to operate at a leading power factor; underexcitation causes it to operate at a lagging power factor. Operating a synchronous motor at no load and greatly overexcited field causes it to take a current which leads the voltage by nearly 90°, like a capacitor. When operated this way, the synchronous motor is often referred to as a *synchronous condenser*. The characteristics will now be discussed.

10.2 *Construction*

The constructional details of a three-phase synchronous motor are essentially the same as that of a three-phase synchronous generator, as discussed in Chapter 5. The three-phase armature winding is on the stator and is wound for the same number of poles as the rotor. The required dc excitation source for the rotor field in a synchronous motor can be provided by similar means as was discussed for synchronous generators. Usually, synchronous motors have slip rings and brushes on the rotor. The sliding contacts are not always wanted or permitted in certain applications; for example, in chemical plants or in "hot zones" in nuclear power stations. They are then built with brushless excitation systems. The rotor of a synchronous motor can be of the salient-pole type or cylindrical-pole type of construction (see Figures 10.1 and 10.2, respectively). Salient-type rotors, however, in addition to the excitation winding, have an added winding in each pole face. It resembles a squirrel-cage winding, and is normally referred to as a *damper winding*; Figure 10.3 shows the details. As shown, this winding is placed in slots located in the pole faces and parallel to the shaft. The ends of the copper bars are short-circuited in the same manner as the squirrel-cage induction motor rotor (Chapter 8). Hence, a synchronous motor equipped with a damper winding will be self-starting, as it starts by induction motor action.

As mentioned, the armature winding is identical to that of the synchronous generator and identical to that of the stator winding of an induction motor as well. Therefore, when a three-phase voltage source is applied to the armature winding, a rotating field of constant magnitude is produced in the air gap. As we may recall from Eq. 8.1, the speed of this rotating field depends on the line frequency f and the number of poles P in the armature winding:

FIGURE 10.1. **Salient-pole rotor and stator of a converter-fed rolling mill motor; rated capacity 4000 kW at 60 to 120 r/min. (Courtesy of Siemens.)**

FIGURE 10.2. Cylindrical rotor of a converter-fed rolling mill motor; rated capacity 2000 kW at 140 to 280 r/min. (Courtesy of Siemens.)

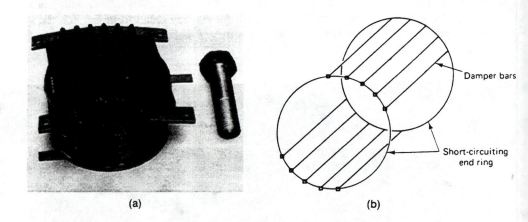

(a)   (b)

FIGURE 10.3. (a) Salient pole with five damper bars short-circuited by end rings; (b) schematic representation of a damper winding for a two-pole rotor. [(a) Courtesy of Siemens.]

$$n_s = \frac{120f}{P} \quad \text{r/min} \tag{10.1}$$

where n_s is the synchronous speed in revolutions per minute. As you may have noticed, unlike the induction motor, a synchronous motor receives its excitaton from two different voltage sources. One is the three-phase ac source to its stator winding and the other is a dc source for the rotor field winding. It is the interaction of these two fields in the air gap that produces the torque, provided those fields revolve at the same speed, that is, synchronous speed.

10.3 Principle of Operation

The important feature of synchronous motors that distinguishes them from other motors is that they need two sources. Let us consider the two-pole synchronous motor in Figure 10.4. A three-phase ac voltage is applied to the stator winding, and a dc source supplies the rotor field. The rotor is assumed stationary. The stator field rotates at synchronous speed, which according to Eq. 10.1 is 3600 r/min for the two-pole machine supplied with a conventional 60-Hz three-phase voltage system. The dc field is stationary because we assumed the rotor is not turning. It can be seen that in order to develop a continuous torque, the two fields must be stationary with respect to each other. As we now appreciate, this occurs only when the rotor is also turning at synchronous speed. It is only then when stator and rotor fields "lock-in," hence the name *synchronous motor*. Figure 10.5 shows this condition; the south pole of the rotor will lock in with the stator north pole, and vice versa. There may be momentary fluctuations in speed, but on the average the speed is constant. If the average speed of the rotor is different from synchronous value, even by a small amount, the poles lose their "grip" and the machine will come to a standstill. The bond between stator and rotor poles is then lost, which is the essential criteria for the development of torque.

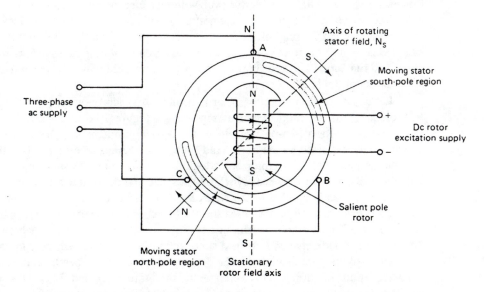

FIGURE 10.4. Instantaneous stator field position with respect to stationary rotor in a synchronous motor.

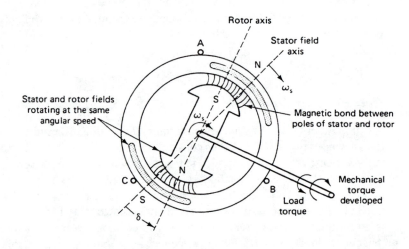

FIGURE 10.5.   Torque development in synchronous motor.

Before we proceed any further, let us give the situation in Figure 10.4 a little thought. We have a pair of revolving armature poles that pass by the stationary rotor poles at great speed. In the position shown there will momentarily be a positive torque developed between the pairs of poles of opposite magnetic polarity, "positive" being defined as the tendency of the rotor north pole to follow the stator south pole. Thus the rotor tends to rotate clockwise. However, because of the rotor inertia, even without a mechanical load on its shaft, the stator field slides by so fast that the rotor cannot follow it. As a result, the rotor does not move, and we say that the starting torque is zero. It is for this reason that the damper winding is included in each of the rotor pole faces. The synchronous motor then starts as an induction motor and runs up to slip speed; at that point the rotor field winding is energized, and the rotor and stator fields will lock in.

Let us now assume that the rotor is turning at synchronous speed as well. Figure 10.5 predicts this situation, which is repeated in Figure 10.6 to compare with the no-load condition. For a two-pole machine, the revolving magnetic poles on the stator are schematically indicated by the north- and south-pole regions (see Figure 10.6a). With no load on the motor shaft the rotor poles lock in with the opposite stator poles, as shown. The whole arrangment rotates at synchronous speed and the torque angle δ is zero. When a mechanical load is applied to the shaft, the rotor tends to slow down somewhat. The mechanism involved will be discussed shortly, but let us assume that at full load the position is as shown in Figure 10.6b. Again, this entire arrangement rotates at synchronous speed. The magnetic bond between rotor and stator field is still maintained, except that the rotor is displaced by the torque angle δ. The developed torque, T_d, depends on the angle δ, and of course it must be sufficient to overcome the shaft torque applied, T_{load}. Next we discuss how the synchronous motor adjusts itself to increased loading.

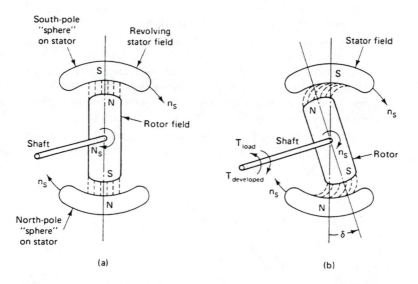

FIGURE 10.6. **Schematic illustration of rotating rotor with respect to rotating stator field with increasing shaft load: (a) no-load condition; (b) full-load condition (note power or torque angle, indicating amount rotor is slipped back).**

10.4   *Loading Synchronous Motors*

Like all motors, the synchronous motor simultaneously acts as a generator as well as a motor during operation. In the dc shunt motor, for example, we have seen that increasing the shaft load reduces the motor speed, and consequently the counter EMF. This results in increased current flow through the armature, to compensate for the increased torque and power by the load. It is the counter EMF which limits the armature current.

Contrary to dc machines, the synchronous motor runs at absolutely constant average speed, regardless of the load. The question then becomes: How does the synchronous motor adjust itself to an increased shaft load? In answering this, the following points should be noted illustrating the fundamental differences in behavior. First, the counter EMF in a synchronous motor does not necessarily have to be smaller than the applied voltage, as it is in a dc motor. In fact, it can be equal to or even larger than the applied voltage. This should not surprise us, we have already encountered a similar situation. As you may recall, in the synchronous generator the generated voltage can be larger or smaller than the terminal voltage. Second, the phase relation between the counter EMF E_c and the terminal voltage V in a synchronous motor is not fixed at $180°$ (i.e., they are not directly opposite). Third, the armature current I_A in a synchronous motor lags behind the resultant voltage by nearly $90°$, while in a dc motor it is always in phase. These points will be clarified shortly.

Going back to our original question of how the synchronous motor adjusts itself to load changes, let us assume the shaft load increases. Since the average speed must remain constant, it cannot draw increased line current in the same manner as the dc motor does (i.e., operate at reduced speed). To see what adjustments do take place in a synchronous motor, we refer to Figure 10.7. Two salient rotor poles are shown at an instant when they are opposite two conductors in the stator. The conductors are assumed to be a pole pitch apart (see Figure 10.7a). The counter EMF induced in the conductors is as shown, being maximum when that conductor is opposite the pole center and zero when midway in between. For any other conductor position with respect to the pole faces, the induced voltage is as shown by curve E_c.

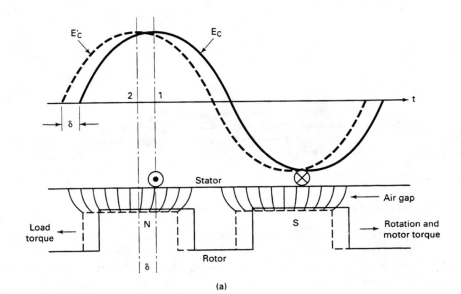

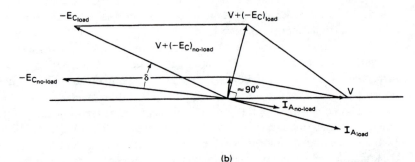

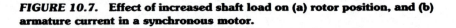

FIGURE 10.7. Effect of increased shaft load on (a) rotor position, and (b) armature current in a synchronous motor.

Upon applying the increased load to the motor shaft, a momentary slowing down of the rotor results, since it takes time for a motor to take increased power from the line. In other words, although still rotating at synchronous speed, the rotor will have slipped back in space as a result of increased loading. This action may be visualized as follows. Assume the magnetic bond between stator and rotor fields as elastic rubber bands. When increasing the shaft load the rubber bands are stretched by an amount dictated by the load torque. This causes the rotor to slip back in space by an angle δ behind the stator field, but it continues to rotate at the same speed provided the rubber bands are intact. The induced EMF corresponding to this new rotor position with respect to the stator field is shown in Figure 10.7a as E_c'. Also, the pole center is shown to be moved from position 1 to position 2 by the same angle δ. This can be illustrated further by the use of a phasor diagram. At light loads, E_c and V are almost directly opposite, while with increasing load the rotor poles slip back by an angle δ. As indicated in Figure 10.7b, with increasing load the phasor E_c slips back from its no-load position by the angle δ. This angle at full load may be as much as 60 electrical degrees for a two-pole machine at full load.

The diagram tells us further that the net voltage $V + (-E_c)$ depends on the position of E_c. The value of I_A is directly proportional to this vector sum. Since the stator winding is predominantly inductive, the current will lag behind the net voltage by an angle of approximately $90°$. In analogy to the dc shunt motor analysis, we can state that the armature current I_A in the synchronous motor is,

$$\mathbf{I}_A = \frac{\mathbf{V} + (-\mathbf{E}_c)}{\mathbf{Z}_s} \tag{10.2}$$

where $\mathbf{Z}_s$ is the impedance per phase of the stator winding. The boldfaced quantities are to remind us that we are considering vector quantities, that is, their proper phase relationships must be accounted for.

Thus by loading the machine the rotor assumes an angular position back from its no-load position. This causes the motor to take increased power from the line to compensate for the increased shaft load, without changing its average speed. The total power supplied to the motor per phase is

$$P = VI_A \cos\theta \quad \text{W/phase} \tag{10.3}$$

where V is the phase voltage in volts and θ is the angle between $\mathbf{V}$ and $\mathbf{I}_A$. The total mechanical power developed (the air gap power) is

$$P_d = E_c I_A \cos\alpha \quad \text{W/phase} \tag{10.4}$$

where α is the phase angle between the vectors $\mathbf{E}_c$ and $\mathbf{I}_A$. The net shaft power is less than P_d by an amount equalling the rotational losses. The difference between P and P_d is the armature copper loss $I_A^2 R_A$.

Finally, the rubber bands will break if the load on the machine continues to be increased. When this happens we speak of the machine losing synchronism. In other words, we expected too much of our motor. The point at which this occurs is generally referred to as the *pullout torque*.

✳ ✳ 10.5 *Adjusting the Field Excitation*

When the dc excitation current is increased, the speed of the synchronous motor remains constant, but its induced EMF E_c must increase because of the strengthened rotor field. It may appear that the motor would stall or start to act as a generator when the induced EMF becomes equal to or greater than the line voltage, respectively. The synchronous motor, however, continues to operate as a motor even though the counter EMF exceeds the line voltage. In this condition the motor is said to operate overexcited, with the result that it takes power from the line with a leading power factor. Figure 10.8 shows us how this is accomplished.

Assume the motor is adjusted such that it is operating at unity power factor, that is, the armature current I_A is in phase with the line voltage, as indicated in Figure 10.8a. As usual, all calculations are made on a per phase basis. From Eq. 10.2 we see that the armature current is proportional to the vector sum of the line voltage $\mathbf{V}$ and the counter EMF E_c, limited by the impedance $\mathbf{Z}_s$. Under our assumption of increased field excitation E_c increases, but the load on the machine has not increased, and the angle δ remains unchanged. Since the vector sum of $\mathbf{V}$ and $\mathbf{E}_c$ increases and the impedance being constant, the armature current must increase. However, the load power remains constant, which implies that the phase angle must change in such a manner that the in-phase component of the armature current with the terminal voltage remains unchanged. That is, $\mathbf{I}_A \cos \theta$ is constant. This can be accomplished only if the armature current assumes a leading position as shown in Figure 10.8b.

Considering next a decrease in the field excitation, the counter EMF $\mathbf{E}_c$ will become smaller. As before, the in-phase component of the armature current with the line voltage $\mathbf{V}$ must remain the same since the power delivered is constant. This means that the vector

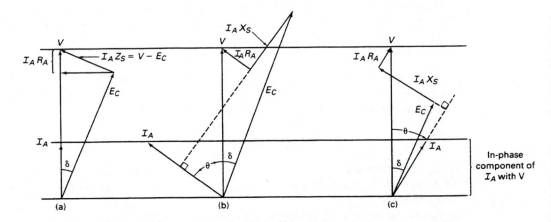

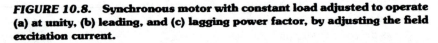

FIGURE 10.8. **Synchronous motor with constant load adjusted to operate (a) at unity, (b) leading, and (c) lagging power factor, by adjusting the field excitation current.**

sum of $\mathbf{V}$ and $\mathbf{E}_c$ again becomes larger as compared to the unity power factor case. The armature current will increase depending on this voltage increase and can do so only by changing its power factor to a lagging angle, as illustrated in Figure 10.8c.

The reader should note the constructional difference of phasor diagram Figure 10.8 as compared to Figure 10.7. In Figure 10.8, $\mathbf{E}_c$ is represented, whereas in Figure 10.7 it was represented as $-\mathbf{E}_c$. Of course, this does not represent anything different; the minus sign associated with the phasor $\mathbf{E}_c$ simply means a 180° phase shift. In both instances, as Eq. 10.2 testifies,

$$\mathbf{I}_A \mathbf{Z}_s = \mathbf{V} + (-\mathbf{E}_c) = \mathbf{V} - \mathbf{E}_c \quad \mathbf{V} \tag{10.5}$$

where $\mathbf{I}_A \; \mathbf{Z}_s$ is the impedance voltage drop in the armature circuit. The synchronous impedance $\mathbf{Z}_s$ of the motor is obtained in similar fashion as that for the synchronous generator. As expected, the electric circuit representation is also identical. Naturally, the power flow is reversed, as we will see shortly. Thus an increase in dc excitation results in a leading power factor (overexcited), while a decrease in field excitation results in a lagging power factor (underexcited), with no appreciable increase in input power to the machine.

✳✳ 10.6 *Phasor Diagrams*

Expanding on the phasor diagram of Figure 10.8a, Figure 10.9a depicts a load condition with the excitation of the field such that unity power factor results. The diagram shows the armature current and voltage drop due to the impedance of the motor. In vector notation,

$$\mathbf{V} = \mathbf{E}_c + \mathbf{I}_A \mathbf{Z}_s \quad \mathbf{V} \tag{10.6}$$

the $I_A R_A$ voltage drop is due to the armature resistance, and the reactive voltage drop $I_A X_s$ is due to the armature reaction and armature reactance effects, similar to that in the synchronous generator. As expected, the equivalent circuit diagram of the synchronous motor, shown in Figure 10.9b, is similar to that of the synchronous generator, except the power flow is reversed since the motor takes power from the line. As we will see, this implies that $\mathbf{V}$ leads $\mathbf{E}_c$ in the phasor diagram. The counter EMF $\mathbf{E}_c$ is a function of the dc excitation current.

If the field excitation is now increased, the counter EMF increases, causing the current to assume a leading power factor. The phasor diagram corresponding to this condition (assuming no change in load) is represented in Figure 10.10a. Figure 10.10b indicates the situation where the synchronous motor is underexcited and the line current lags the applied voltage by an angle θ. Let us compare the phasor diagrams for the three kinds of excitation, that is, unity power factor as depicted in Figure 10.9a, overexcited and underexcited as in Figure 10.10. It is apparent that with constant torque developed by the motor as expressed by keeping the torque angle δ the same, the phase angle θ can be adjusted from leading to lagging or, vice versa, by merely changing the dc field excitation current. This in turn modifies the counter EMF as explained before, which as can be seen, rotates the voltage-drop triangle formed by the $I_A R_A$, $I_A X_s$, and $I_A Z_s$ vectors. When the phase angle departs from unity power factor, the size of this triangle increases, reflecting an increase in the line

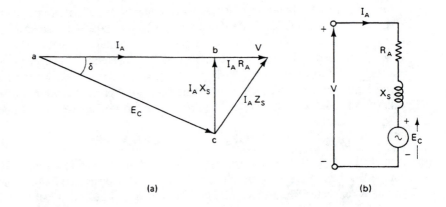

(a)

(b)

FIGURE 10.9. (a) Phasor diagram of a synchronous motor (per phase) operating at unity power factor; (b) equivalent circuit diagram.

current. Because of the constant torque delivered by the machine, the in-phase component of the current with the voltage remains constant.

In reality the power angle is not quite fixed, even for constant loads, but to some extent depends on the field current as well. For an increased field current the rotor pole field

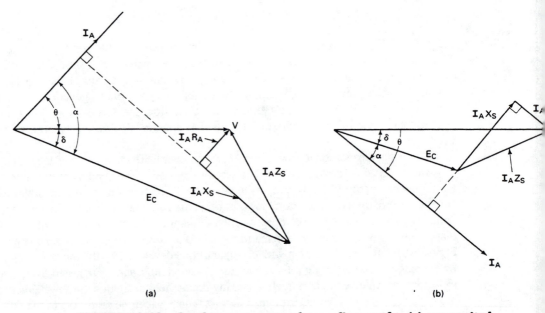

(a)

(b)

FIGURE 10.10. Synchronous motor phasor diagram for (a) overexcited and (b) underexcited conditions.

becomes stronger; therefore, less slippage occurs between the two fields with increasing load.

EXAMPLE 10.1

A 75-hp 440-V 900- r/min three-phase Y-connected synchronous motor has an effective armature resistance of 0.15 Ω and a synchronous reactance of 2.0 Ω/phase. At rated load and PF = 0.8 leading, determine:

a. the torque angle δ, and

b. the developed mechanical power P_d.

Assume the motor has an efficiency (excluding the field loss) of 90%.

Solution The motor input power is

$$P_{in} = \frac{hp \times 746}{\eta} = \frac{75 \times 746}{0.90} = 62,167 \text{ W}$$

the line current

$$I = \frac{P_{in}}{\sqrt{3}V_L \cos\theta} = \frac{62,167}{\sqrt{3} \times 440 \times 0.8} = 102 \text{ A}$$

and the voltage per phase is

$$\frac{440}{\sqrt{3}} = 254 \text{ V}$$

Therefore,

$$\theta = \cos^{-1} 0.8 = 36.9° \text{ leading}$$

a. From Figure 10.10a, taking the terminal voltage V as a reference, and $Z_S = \sqrt{2^2 + 0.15^2}\angle \tan^{-1} 2/0.15 = 2.006\angle 85.7°\Omega$ E_c is determined as

$$
\begin{aligned}
\mathbf{E}_c = \mathbf{V} - \mathbf{I}_A\mathbf{Z}_s &= V\angle 0° - I_A\angle\theta \times Z_S\angle\beta \\
&= 254\angle 0° - 102\angle 36.9° \times 2.006\angle 85.7 \\
&= 364 - j172 = 403\angle -25.3° \text{ V}
\end{aligned}
$$

The negative sign indicates that $\mathbf{E}_c$ lags $\mathbf{V}$ by 25.3°, which also represents the torque angle δ.

b. The mechanical power developed follows from Eq. 10.4, namely,

$$
\begin{aligned}
P_d &= 3E_c I_A \cos\alpha \\
&= 3 \times 403 \times 102 \times \cos 62.2° = 57,484 \text{ W}
\end{aligned}
$$

which is also equal to the power input minus the armature resistance loss, or

$$
\begin{aligned}
P_d &= P_{in} - 3I_A^2 R_A \\
&= 62,167 - 3 \times 102^2 \times 0.15 = 57,484 \text{ W}
\end{aligned}
$$

Note that, as before, the factor of 3 arises from the fact that there are three phases. The power developed at the motor shaft is less than P_d by an amount equal to the rotational losses. A simple calculation shows that this amounts to $P_d - P_{out} = 1534$ W for this motor.

From the phasor diagrams it is seen that $\mathbf{E}_c$ is lagging the terminal voltage consistent with the calculated results. This is opposite to synchronous generators, where $\mathbf{E}_c$ leads $\mathbf{V}$. Let us examine this in some more detail.

✳ ✴ 10.7   *Power Flow*

As can be seen from the phasor diagrams for the synchronous motor (e.g., Figure 10.10), the terminal voltage $\mathbf{V}$ of the machine is leading the counter EMF $\mathbf{E}_c$. It is independent on the power factor at which the motor is operating. The reader may remember from our study on synchronous generators that the terminal voltage is constantly lagging the generated EMF. Again this is independent on the power factor, only the amount by which it is leading or lagging depends on the power factor. We know the power factor can be controlled in both machines by adjustment of the dc field excitation current.

We can now appreciate what is taking place. In a synchronous generator, when the prime-mover torque is increased, it tends to speed up the generator away from the system bus. We have seen that it is locked onto the bus and rather than pulling away, the rotor advances a certain angle with respect to the resultant air gap field. The generated EMF waveform will follow this rotor advancement. This is shown in Figure 10.11. The angle δ, the torque or power angle, is a measure of the real power delivered by the machine. It is positive when $\mathbf{E}_G$ leads $\mathbf{V}$, that is, power flows out of the machine. If we apply a negative torque, by lowering our prime-mover torque from that value required by the load, the network would than pull our machine. Another way of putting this is to say that it acts as a motor—the machine will pull some of the load. In that case $\mathbf{E}_c$ would lag $\mathbf{V}$ and the resultant air gap flux would now start to lag the rotor position. This implies power flowing into the machine. Figure 10.11 predicts this action where $\delta < 0$.

As we change the field current in either operating condition, no change can take place in the real power or torque delivered by the prime mover. An indirect change will take place in an actual machine, amounting to a slight adjustment of the power angle, as we have discussed. However, to simplify matters we will ignore this change. Figure 10.12 summarizes the important differences between generator and motor action.

Consider now the situation where the torque on the motor is being increased. Under this assumption the developed power will reach a maximum when δ equals $-90°$ for a motor or $+90°$ for a generator, as discussed in Section 6.7. Figure 10.12c shows the developed power P as a function of δ. This curve, as we have seen, follows from Eq. 6.31, where it was assumed that the armature resistance is negligible. Increasing the torque beyond this point will result in a loss of synchronism, and the motor will stall. From Eq. 6.31 we have

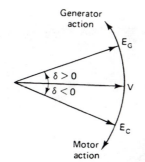

FIGURE 10.11.   Power flow in synchronous machine ($\delta > 0$ generator action, $\delta < 0$ motor action).

$$P = \frac{E_c V}{X_s} \sin \delta \quad \text{W/phase} \qquad (10.7)$$

The maximum power or pullout power, P_{max}, occurs when $\sin \delta = 1.0$ ($\delta = 90°$), or

$$P_{max} = \frac{E_c V}{X_s} \quad \text{W/phase} \qquad (10.8)$$

under the assumption that the armature resistance is negligible. Losing synchronism implies that the magnetic bond between stator and rotor fields is lost. The rotor will "skip" poles and tries to run as an induction motor if the starting winding can support this load. It will continue to do so until the excessive load is removed. In the meantime excessively large currents will flow in the stator winding, which goes hand in hand with dangerously high heating effects, which must be avoided.

EXAMPLE 10.2

A 2200-V three-phase Y-connected synchronous motor has a synchronous reactance $X_s = 2.6$ Ω/phase. The armature resistance is assumed negligible. The input power is 820 kW, while the field excitation is such that the counter EMF is 2800 V. Calculate

a. the torque angle, and

b. the line current and power factor.

Solution   First we construct a phasor diagram to see what is "happening." From Figure 10.12b we see that for the synchronous motor **V** leads $\mathbf{E}_c$; furthermore, $E_c > V$, resulting in a leading PF (see Figure 10.13). The circuit is Y-connected; thus

$$V = \frac{2200}{\sqrt{3}} = 1270 \ \text{V/phase}$$

and

$$E_c = \frac{2800}{\sqrt{3}} = 1617 \ \text{V/phase}$$

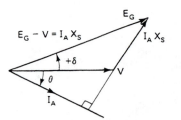

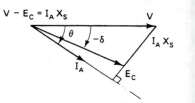

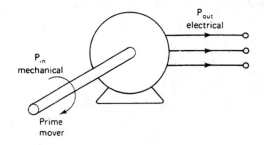

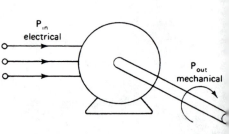

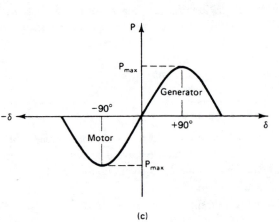

FIGURE 10.12.   Motor and generator action of a synchronous machine (R_A is assumed negligible): (a) generator action; (b) motor action; (c) developed power versus torque (or power) angle δ.

a. From Eq. 10.7,

$$P = \frac{E_c V}{X_s} \sin \delta \quad \text{W/phase}$$

Therefore,

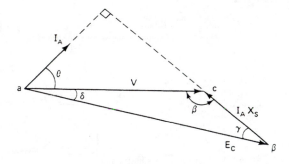

FIGURE 10.13.   Phasor diagram for solution of Example 10.2.

$$\frac{820,000}{3} = \frac{1617 \times 1270}{2.6} \times \sin \delta$$

or $\sin \delta = 0.346$ and $\delta = \sin^{-1} 0.346 = 20.2$ electrical degrees

b. For the PF calculation we need the angle θ. This angle is obtained from Eq. 10.2, as

$$I_A \angle \theta = \frac{V \angle 0° - E_c \angle - \delta}{j X_s}$$

$$= \frac{1270 \angle 0° - 1617 \angle -20.2}{j2.6} = 235 \angle 24°$$

Hence $\theta = 24°$ and the power factor PF $= \cos 24° = 0.914$ (leading).

✳ ✳ 10.8   *Efficiency*

To complete our picture of the power flow, we must consider the losses occurring in a synchronous motor. As expected, this is similar to the synchronous generator. Figure 10.14 shows the losses occurring in a synchronous motor. The motor losses subtracted from the electrical input give efficiencies generally between 80% and the high 90s. Losses appear as heat that must be removed. As we see again, efficiency figures can vary widely, depending on who is figuring: some do not include auxiliary needs, such as exciter losses.

To calculate the efficiency, we proceed as before, namely,

$$\eta = \frac{P_{out}}{P_{out} + \text{losses}} \times 100\% \tag{10.9}$$

where P_{out} is the output shaft power, which equals the shaft torque times the radial speed at which it is developed,

$$P_{out} = \omega T_{shaft} \quad \text{W} \tag{10.10}$$

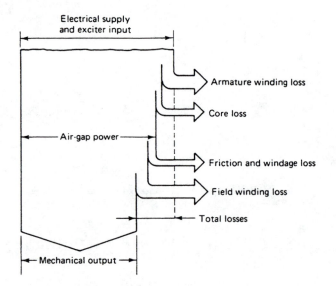

FIGURE 10.14. **Power flow for a synchronous motor.**

The power input is simply

$$P_{in} = \sqrt{3}V_{L-L}I_L \cos\theta \qquad (10.11)$$

so that Eq. 10.9 can also be written as

$$\eta = \frac{\omega T_{shaft}}{\omega T_{shaft} + 3I_A^2 + I_F V_F + P_{core} + P_{fr+w}} \times 100\% \qquad (10.12)$$

with the various symbols representing the quantities discussed before.

Following the method discussed under synchronous generators, we will normally exclude the auxiliary losses. In that case, P_{out} becomes

$$P_{out} = P_d - P_{fr+w+core} = P_d - P_{rot} \quad \text{W} \qquad (10.13)$$

from which the shaft torque (in N.m) can be established from Eq. 10.10.

EXAMPLE 10.3

A 415-V eight-pole 60-Hz Δ-connected three-phase synchronous motor has its field excitation adjusted, resulting in an induced EMF of 520 V and a torque angle of 12 electrical degrees. The armature impedance $Z_s = 0.5 + j4.0$ Ω. The rotational losses amount to 2000 W. Determine

a. the line current,

b. the power factor,

c. the output power,

d. the efficiency, and

e. the shaft horsepower and torque.

Solution   The phasor diagram for the given values is shown in Figure 10.15. From Eq. 10.2 we have (or by inspection of the phasor diagram)

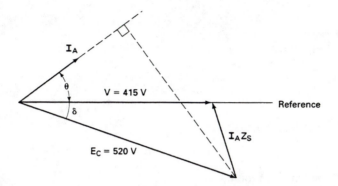

FIGURE 10.15.   Phasor diagram for the solution of Example 10.3 (on a per phase basis).

$$I_A Z_s = V\angle0° - E_c\angle - \delta = 415 - 520\angle - 12°$$
$$= 415 - (508.6 - j108.1) = 143.0\angle130.9° \text{ V}$$

Since

$$Z_s = 0.5 + j4.0 = 4.031\angle82.9° \ \Omega$$

there results

$$I_A = \frac{I_A Z_s}{Z_s} = \frac{143\angle130.9°}{4.031\angle82.9°} = 35.5\angle48° \text{ A}$$

a. Therefore,

$$I_L = I_A\sqrt{3} = 35\sqrt{3} = 61.5 \text{ A}   \text{ (armature } \Delta-\text{connected)}$$

b. $\theta = 48°$ and PF $= \cos 48° = 0.669$ leading.

c. The input power can be calculated:

$$P_{in} = \sqrt{3}V_{L-L}I_L \cos\theta = 3V I_A \cos\theta$$
$$= 3 \times 415 \times 35.5 \times 0.669 = 29,573 \text{ W}$$

The armature copper loss is

$$P_{cu} = 3I_A^2 R_A = 3 \times 35.5^2 \times 0.5 = 1890 \text{ W}$$

The developed air gap power is then

$$P_d = P_{in} - P_{cu} = 3E_c I_A \cos(\theta + \delta)$$
$$= 29,573 - 1890 = 27,683 \text{ W}$$

The output power is

$$P_{out} = P_d - P_{rot} = 27,683 - 2000 = 25,683 \text{ W}$$

d. The efficiency is

$$\eta = \frac{P_{out}}{P_{in}} \times 100 = \frac{25,683}{29,573} \times 100 = 86.8\%$$

e. The shaft horsepower is

$$hp = \frac{25,683}{746} = 34.4$$

and the load torque is

$$T_{shaft} = \frac{25,683}{2\pi \times 900/60} = 273 \text{ N·m}$$

10.9 Synchronous Motor V Curves

As has been demonstrated, the power factor of a synchronous motor can be controlled by variation of the field current. We have seen from the examples that not only does the power factor change, but the line current is affected as well. Under the assumption of a constant load, the power input to the motor does not change appreciably and can be assumed to remain the same. Of course, when the current increases, the copper losses increase somewhat, as does the core loss due to a slight change in flux. However, these increased losses do not change the input power too much.

To show how the field current controls the power factor, consider a large reduction in field excitation, which will reduce the induced EMF E_c and cause the line current to lag. Also, we assume the motor to operate at no load. Increasing the field current from this small value, we see that the line current decreases until a minimum line current occurs, indicating that the motor is operating at unity power factor. Up to this point the motor was operating at a lagging power factor. Continuing to increase the field current, the line current increases again and the motor starts to operate at a leading power factor. Plotting the relationship of armature current versus field excitation current, the lowest curve in Figure 10.16 is obtained. Repeating this procedure at various increased motor loads results in a family of curves, as shown.

Because of their resulting shape, these curves are commonly referred to as *V curves*. The point at which unity power factor occurs is at the point where the armature current is

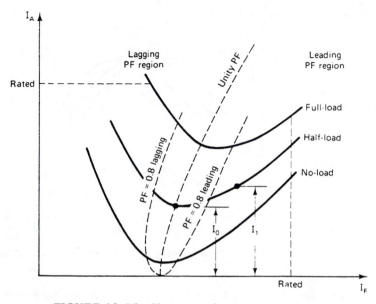

FIGURE 10.16.   V curves of a synchronous motor.

minimum. This is indicated by I_0, for example, for the curve taken when the synchronous motor delivers 50% of its rated load (see Figure 10.16). Connecting the lowest points of all V curves results in the dashed curve indicated by "unity PF" in the figure. This curve is commonly called a *compounding curve*. Others can be drawn for different power factors, as shown.

Taking the necessary data during these tests, the power factor may also be plotted for each point. Plotting the power factor versus field current results in a family of curves shown in Figure 10.17, which are inverted V curves. The highest point on each of these curves indicate unity power factor. It shows that if the excitation is adjusted at a certain load for unity power, increasing that load results in a lagging power factor.

These PF curves can also be constructed from the V curves. Referring again to the 50% rated load curve in Figure 10.16, let I_1 be a value of armature current at some power factor θ. The power per phase is

$$P = V I_1 \cos \theta = \frac{1}{2} - load$$

and

$$I_1 \cos \theta = I_0$$

for all values of θ, since the power delivered by the machine for this curve is constant. The excitation corresponding to the value of armature current I_0 is called the normal excitation. Reducing this excitation, the motor takes a lagging current and is said to be *underexcited*, as we have discussed. Increasing the excitation results in a leading current and is said to

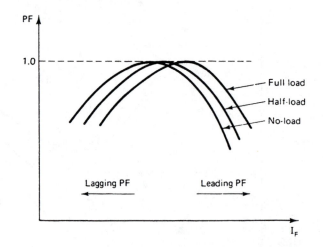

FIGURE 10.17.   **Power factor versus field current at different loads.**

be *overexcited*. Thus the power factor cos $\theta = I_0/I_1$ can therefore be found for any value of armature current I_1 and given power. The corresponding field current I_f at the selected values of armature current can be read from the curve, providing the necessary data to plot the curves in Figure 10.17.

✳ ✳ 10.10   *Synchronous Condensers*

When a synchronous motor is operated without load, it still takes some power from the line to overcome its internal losses, such as friction and windage, and core losses. As we have seen before, these losses are minimal and therefore the in-phase component of the armature current is rather small, although the current may be large, depending on the dc excitation. If the motor is operated overexcited it will take a very low leading power factor current and behave like a capacitor. This can be verified by referring to Figure 10.18, which shows the armature current to lead the supply voltage by nearly 90°.

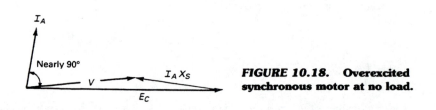

FIGURE 10.18.   **Overexcited synchronous motor at no load.**

     In large plants with poor power factors, this feature is taken advantage of. The synchronous machine is then placed in parallel at the incoming power lines to the plant.

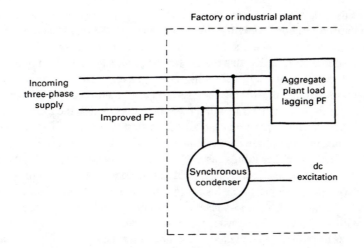

FIGURE 10.19.   Power factor improvement by using a synchronous condenser.

Hence the overall power factor of the incoming line to the plant can be improved. Such a connection is indicated schematically in Figure 10.19. When a synchronous motor is employed this way, it is commonly referred to as a *synchronous condenser*. It will operate at no load but with a strong dc field excitation.

10.11   *Power Factor Correction in Industrial Plants*

When an industrial plant buys power from the electric utility, the total amount of electrical energy is measured as well as the maximum rate at which this energy is consumed. These measurements (or readings) are normally taken on a monthly basis, and a portion of the bill reflects the rate of consumption or *demand charge*, as it is called.

The demand charge stems from the way the plants use this energy, that is, it relates directly to the resistive component and the reactive component. Only the resistive component provides useful work, but the reactive component is a necessary ingredient in most machines as we realize, since it makes things happen. The relationship between the two components is of course the power factor.

As we have seen in our study of the various machines, the reactive power needed is a function of the type of motor and transformer installed. For example, lightly loaded induction motors (as well as transformers) have poor power factors. This implies that the reactive component is relatively large. There is not much we can do about this, but it is possible to take some corrective action. That is where capacitive action, in particular our synchronous motor, comes in. When the plant takes this corrective action, we usually refer to it as *power factor correction*. Figure 10.19 demonstrated this in principle, and shortly we will see how we can determine what is needed. Before going into these details, let us consider what benefits are attained. Naturally, it saves energy, which translates into a lower monthly bill; otherwise, why bother? In addition, it can reduce the peak demand and thus add capacity to our supply transformer or even to the plant feeders. An additional benefit may even be a somewhat increased voltage. The power factor is, in principle, best corrected at the source; practically, this means for the plant site at the billing meters.

A convenient way of showing power factor correction applications is by means of solving problems, illustrating the use of vector diagrams. We demonstrate this in the following examples. In dealing with power factor correction the reader may notice that we use kVA and kW rather than volts and amperes. Furthermore, we deal with total quantities now rather than on a per phase basis. This is following standard practical calculations as applied in the "field"; it makes things easier. In the end we want to specify what rating our machine should have to do the job. This should not be confused with the previous calculations that we performed on a per phase basis. When dealing with equivalent circuits and associated phasor diagrams, using one-phase power only simplifies our work considerably. Naturally, knowing the three-phase power, the power factor, and the voltage level we are working at provides us with the knowledge of the line currents.

EXAMPLE 10.4

An industrial plant represents a load of 1600 kVA at a lagging power factor of 0.6. Determine

a. the kVA rating of the synchronous condenser to be installed in parallel with this load to raise the line power factor to unity, and

b. the total kW load.

Solution The overall circuit diagram is as depicted in Figure 10.19, and the corresponding vector diagram is represented in Figure 10.20.

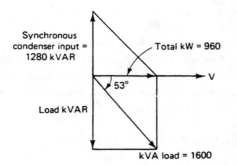

FIGURE 10.20. Vector diagram for solution of Example 10.4.

a. For an overall plant power factor of unity, the synchronous condenser will have to counteract the vertical component (reactive power) of the plant load, which is

$$1600 \times \sin 53° = 1280 \text{ kVAR}$$

b. The total load is

$$1600 \times \cos 53° = 960 \text{ kW}$$

We see that the supply rating is reduced from 1600 kVA to 900 kVA (PF = 1.0) at the expense of a 1280-kVA synchronous condenser. In practice it is seldom necessary to raise the power factor above approximately 85%, since very little is gained beyond this in reducing the line current. The expense of the additional synchronous condenser rating does not warrant it. This may be indicated by the following example.

EXAMPLE 10.5

Referring to the plant data in Example 10.4, determine:
a. the kVA rating of the synchronous condenser required to raise the power factor to 0.85 lagging, and
b. the total kVA load.

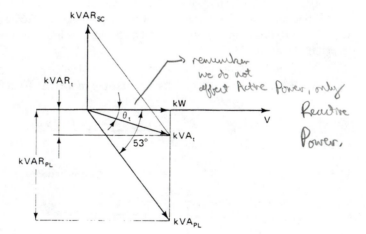

FIGURE 10.21.   Vector diagram for Example 10.5.

Solution

$$\theta_t = \cos^{-1} 0.85 = 31.8°$$

$$\sin \theta_t = 0.527 \quad \text{and} \quad \tan \theta_t = 0.620$$

The vector diagram is as shown in Figure 10.21. From the diagram we see that the new or total kVAR$_t$ is the difference between the kVAR of the plant and that of the synchronous condenser. Therefore,

$$\text{kVAR}_{SC} = \text{kVAR}_{PL} - \text{kVAR}_t$$

Now $\mathrm{kVAR}_{PL} = 1600 \times \sin 53° = 1280$ kVA and since the total power does not change,

$$\mathrm{kVAR}_t = \mathrm{kW} \times \tan \theta_t = 960 \times 0.620 = 595$$

and

$$\mathrm{kVAR}_{SC} = 1280 - 595 = 685$$

b. $\mathrm{kVA}_t = \mathrm{kW}_t / PF = 960 / 0.85 = 1129.$

Comparing Examples 10.4 and 10.5 it is evident that the size of the synchronous condenser is considerably less, namely, 685 kVA as compared to 1280 kVA. This represents a reduction in capacity of approximately 46%, while at the same time the line current is only 1/0.85, or about 18% as large as that compared to unity power factor.

EXAMPLE 10.6

As a further example we can illustrate the power factor improvement calculation if in addition to providing power factor improvement the synchronous condenser is supplying a load. To do so let us assume that we have a system as ilustrated in Figure 10.22, showing a line diagram of a transformer supplying an existing 1600 kVA, 0.6 power factor lagging load. To this plant we wish to add a 750-hp synchronous motor with 0.8 PF leading and an efficiency of 90%.

a. Will this added load overload the transformer?

b. What will be the power factor after the motor has been added?

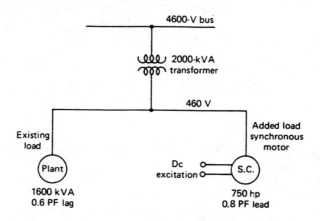

FIGURE 10.22. Line diagram of supply line to plant.

Solution With reference to Figure 10.23, which shows the vector diagram, triangles OAB, BCD, and ODE are right-angle triangles, so that simple trigonometric relations apply. In triangle OAB, being the existing plant load, has the sides

$$OA = 1600 \times \cos 53° = 960 \text{ kW}$$

and
$$EC = AB = 1600 \times \sin 53° = 1280 \ \text{kVA}$$

The new motor which is added has a kW input (vector $BC = AE$) of

$$AE = \frac{750 \times 0.746}{0.90} = 622 \ \text{kW}$$

The leading reactive component (vector DC) when the motor operates at full load and excitation is

$$DC = BC \times \tan 37° = 466.5 \ \text{kVAR}$$

This enables calculation of the new total kVA as

$$\text{kVAR}_t = \sqrt{(OA + AE)^2 + (CE - DC)^2}$$
$$= \sqrt{1582^2 + 813^2} = 1779 \ \text{(vector } OD)$$

and the overall power factor

$$\text{PF}_t = \frac{1582}{1779} = 0.89$$

which is a considerable improvement in power factor. Furthermore, with the added synchronous motor load the line current increases approximately 10%, still within the rating of the transformer. It is interesting to note, however, that if there had been an 750-hp induction motor added instead, the line current would have exceeded the transformer rated current by about 18%. The calculation of this has been deferred to Problem 10.7.

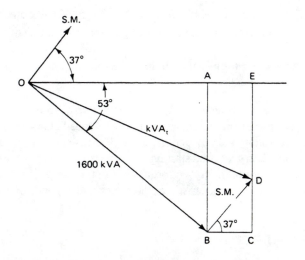

FIGURE 10.23.   Vector diagram for Example 10.6.

10.12 *Applications*

As we have discussed, synchronous motors offer some definite advantages over the motors discussed previously. These include constant-speed operation, power factor control, and high operating efficiencies. The high-efficiency operation relates directly to the synchronous motor having a separate source for the required field excitation. We may well remember that induction machines must have a small air gap in the magnetic circuit. This will limit the magnetizing current needed to establish the flux. The air gap is made as small as mechanical clearance will allow, generally in the range 0.02 to 0.05 in. Induction machines are excited from a single source, and the excitation current and power component of the current flow in the same lines. The magnetizing current is a large component of the rated current, so that they are found to operate at relatively poor power factors at light loads.

Since the field has its own supply, synchronous motors usually have larger air gaps than those found on induction motors. In addition, as we have seen, they can operate at unity power factor, resulting in efficiencies for this motor in the range 85 to 97%. As such, they are the most efficient of all motors.

Today, the starting and running characteristics of synchronous motors are such that they can successfully compete with induction motors. Because the synchronous motor needs a dc source and slip rings, the initial cost of the machine is larger than that of a comparably sized induction motor, particularly for small units. In the larger sizes for heavy industrial services, this disadvantage no longer exists. As a matter of fact, there is a horsepower range and speed range which puts the synchronous motor at an advantage. It is precisely for these reasons that synchronous motor applications are increasing.

When considering a synchronous motor for a specific service, we need to concern ourselves basically with the following specifications.

1. The starting torque, representing the ability of the motor to start the load.
2. The pull-in torque, the ability to pull into synchronism from induction to synchronize operation.
3. The pullout torque, the ability to maintain synchronism under full-load condition or pulsating loads.

As discussed previously, the first two characteristics are associated with the damper winding. It is for this reason that the starting torque-speed characteristic of a synchronous motor is identical to that of an induction motor. The resistance of the damper winding can be made high enough to give good starting characteristics (typically ranging from about 110 to 170% of full-load torque). Pull-in torques are in the range of 120%. The pullout torque, on the other hand, is a characteristic of the motor and depends on the field strengths.

PROBLEMS

10.1   Electrical power is to be supplied to a three-phase 25-Hz system from a three-phase 60-Hz system through a motor-generator set consisting of two directly coupled synchronous machines.

a. What is the minimum number of poles the motor may have?

b. What is the minimum number of poles the generator may have?

c. At what speed in revolutions per minute will the set operate?

10.2 A 60-Hz synchronous motor is coupled to, and drives, a 50-Hz synchronous generator. How many poles does each machine have, and at what speed does the motor-generator set operate?

10.3

a. What is the synchronous speed of a 12-pole 460-V 60-Hz synchronous motor?

b. At what speed will it run when the full load is reduced to 50%?

c. What is the speed if the supply voltage is 380 V?

10.4 A three-phase 6600-V Y-connected synchronous motor was tested. The open-circuit and short-circuit data on this machine are as follows:

Open circuit:

I_F (A)	40	80	120	160	240
V_{L-L} (V)	3800	5800	7000	7800	8700

Short circuit:

$$I_F = 180 \text{ A}, \quad I_A = 460 \text{ A(rated current)}$$

When running as a synchronous motor it takes 80 A at a power factor of 0.6 leading. What field current is necessary? Assume that R_A is negligible.

10.5 The motor of Problem 10.4 has 12 poles and a field circuit resistance of 0.2 Ω. If the friction and windage losses are 5.65 kW and the magnetic losses equivalent to 4.0 kW; at the given load:

a. What is the efficiency of the motor?

b. What is the shaft horsepower developed?

10.6 A 2000-V three-phase Y-connected synchronous motor has a synchronous reactance of 2.2 Ω/phase. The input is 800 kW at normal voltage and the induced line EMF is 2500 V. Calculate the line current and power factor. Assume R_A is negligible.

10.7 In Example 10.6 assume that a 750-hp induction motor was added (otherwise same data) to the system instead of the synchronous motor. Would this overload the transformer? If so, what is the transformer current in this case?

10.8 A 1500-hp 6600-V three-phase Y-connected synchronous motor with a synchronous reactance of 6 Ω/phase delivers full load at unity power factor, and normal rated voltage. The armature resistance is negligible. What percentage increase in field excitation current is needed to achieve PF = 0.866 leading? Assume a linear operation and constant torque angle.

10.9 A three-phase 220-V Y-connected synchronous motor operates with an excitation voltage of 156 V per phase. The armature current is 120 A at unity power factor. Determine the synchronous reactance for this machine, assuming R_A is negligible.

10.10 The input to a 13,800-V three-phase Y-connected synchronous motor is 75 A. The synchronous reactance is 25 Ω per phase. The armature resistance may be assumed negligible.

    a. Find the power supplied to the motor, and the induced EMF for a PF = 0.8 leading.

    b. For a PF = 0.8 lagging, what then is the induced EMF, the line current, and the power supplied to the motor? Assume that the torque angle remains constant at the value determined in part (a).

10.11 A synchronous motor is operating at half-load. An increase in its field excitation causes a decrease in the line current. Does the armature current lead or lag the terminal voltage?

10.12 A synchronous motor delivers full load with the excitation adjusted so that the line current is at its minimum value. If the load is reduced to 50% of its full-load value, explain what effect it will have on the power factor.

10.13 A synchronous motor has a Δ-connected armature with a synchronous reactance of 2.4 Ω/phase. If this motor takes a line current of 68 A at 220 V, with the field adjusted to give an excitation voltage of 160 V/phase:

    a. What power does the motor take from the line?

    b. At what power factor is it operating?

Assume the armature resistance to be negligible.

10.14 A three-phase Y-connected synchronous motor has negligible armature resistance and a synchronous reactance of 12 Ω/phase. The motor takes 12,000 kW at a line voltage of 13,200 V. If the field current is adjusted so that the excitation voltage is 9000 V/phase:

    a. What is the torque angle?

    b. At what power factor is the motor operating?

    c. What is its armature current?

10.15 A 30-hp 440-V Y-connected synchronous motor is operating at a load that causes the rotor to lag the stator field by 10 electrical degrees. If the field excitation is adjusted to produce a generated phase voltage equal to the applied voltage:

    a. What is the resultant voltage per phase in the armature?

    b. What angle does it make with the applied voltage?

10.16 A three-phase 100-hp induction motor has an efficiency of 96% and is operating at a power factor 0.92 lagging. It is in parallel with a three-phase synchronous motor taking 200 kVA at 0.6 power factor leading. The supply voltage is 2400 V. Calculate the line current and power factor of this combined load.

10.17 An industrial factory has a load of 1600 kVA at an average power factor of 0.6 lagging. Calculate

    a. the kVA input to a synchronous condenser to obtain an overall power factor of 0.88 lagging, and

b. the total kW load.

10.18 A manufacturing plant has a load of 3600 kVA at a lagging power factor of 0.707. A 500-hp synchronous motor having an efficiency of 90% is installed and is used to improve the overall power factor to 0.90 lagging.

    a. Draw the kVA vector diagram to represent these conditions. Label each vector clearly.

    b. Calculate the kilovolt-ampere input rating of the synchronous motor and the power factor at which it must operate.

Units

The International System of Units (abbreviated SI, for Système International d'Unités), which is used in this book, is now standard in electrical engineering and much of physics. The SI base units for the fundamental dimensions of length, mass, time, electric current, temperature, and luminous intensity are the meter (m), kilogram (kg), second (s), ampere (A), kelvin (K), and candela (cd), respectively. The dimensions of other quantities are expressed in terms of those base units.

Table A.1 lists the prefixes used with any of the SI units, their abbreviations, and the power of 10 each represents. Table A.2 gives the conversion factors between commonly encountered SI units and Imperial units. Table A.3 presents an overview of the symbols used in this book in the order in which they appear in the text, together with their units.

TABLE A.1. **Prefixes used with SI units.**

Prefix	SI symbol	Meaning		
pico	p	0.000 000 000 001	×	10^{-12}
nano	n	0.000 000 001	×	10^{-9}
micro	μ	0.000 001	×	10^{-6}
milli	m	0.001	×	10^{-3}
centi	c	0.01	×	10^{-2}
deci	d	0.1	×	10^{-1}
dekta	da	10	×	10
hecto	h	100	×	10^2
kilo	k	1000	×	10^3
mega	M	1 000 000	×	10^6
giga	G	1 000 000 000	×	10^9
tera	T	1 000 000 000 000	×	10^{12}

TABLE A.2. **Conversion factors.**

	SI unit	Imperial to SI	SI to Imperial
Length	meter (m)	1 inch = 2.54×10^{-2} m 1 foot = 0.305 m 1 yard = 0.914 m 1 mile = 1.6093 km	1 m = 39.37 inches = 3.281 feet = 1.094 yards
Mass	kilogram (kg)	1 ounce = 28.35×10^{-3} kg 1 pound = 0.454 kg	1 kg = 35.27 ounces = 2.205 pounds
Area	meter² (m²)	1 in.² = 6.45×10^{-4} m² 1 ft² = 0.093 m²	1 m² = 1550 in.² = 10.76 ft²
Volume	meter³ (m³)	1 in.³ = 16.3×10^{-6} m³ 1 ft³ = 0.028 m³	1 m³ = 6.102×10^4 in.³ = 35.3 ft³
Force	newton (N)	1 oz (f) = 0.278 N 1 lb (f) = 4.448 N	1 N = 3.597 oz (f) = 0.225 lb (f)
Torque	newton-meter (N·m)	1 lb·ft = 1.356 N·m 1 oz·in. = 7.062×10^{-3} N·m 1 lb·in. = 0.113 N·m	1 N·m = 0.738 lb·ft = 8.851 lb·in. = 141.61 oz·in.
Energy	joule (J)	1 Btu = 1054 J = 0.252 kcal 1 kWh = 3.6×10^6 J 1 ft·lb (5) = 1.356 J 1 W·s = 1 J	1 J = 9.478×10^{-4} Btu = 2.778×10^7 kWh = 1 W·s = 2.389×10^{-4} kcal
Power	watt (W)	1 hp = 746 W = 550 ft·lb (f)/s 1 W = J/s	1 W = 1.341×10^{-3} hp
Magnetic flux	weber (Wb)	1 line = 10^{-8} Wb = 10^{-8} maxwell	1 Wb = 10^8 lines = 10^8 Mx
Magnetic flux density	tesla (T)	1 line/in.² = 1.55×10^{-5} T 1 gauss = 10^{-4} T	1 T = 6.452×10^4 lines/in.² = 10^4 gauss
Magnetizing force	ampere-turns/ meter (At/m)	1 At/in. = 39.37 At/m 1 oersted = 79.578 At/m	1 At/m = 0.0254 At/in. = 0.0126 oersted

TABLE A.3.   **Common Symbols (in the order in which they appear in the text.**

Symbol	Unit	Unit	Abbreviation
ϕ	Magnetic flux	weber	Wb
B	Magnetic flux density	tesla	T
A	Area	meter^2	m^2
I	Electric current	ampere	A
$\mathcal{F}$	Magnetomotive force, MMF	ampere-turn	At
N	Number of turns		
H	Magnetic field intensity	ampere-turn/meter	At/m
$\mathcal{R}$	Reluctance	ampere-turn/weber	At/Wb
l	Length	meter	m
μ	Permeability	henry/meter	H/m
P	Power	watt	W
V	Volume	meter^3	m^3
f	Frequency	hertz	Hz
t	Lamination thickness	meter	m
F	Force	newton	N
e	Electromotive force, EMF	volt	V
v	Velocity	meter/second	m/s
θ	Phase angle	degree	°
E	Generated EMF (rms)	volt	V
t	Time	second	s
a	Parallel paths in armature		
P	Number of poles		
n	Rotational speed	revolution/minute	r/min
z	Conductors in armature		
k	Constant		
V	Voltage	volt	V
R	Resistance	ohm	Ω
T	Torque	newton·meter	N·m
ω	Angular velocity	radian/second	rad/s
r	Radian length	meter	m
η	Efficiency		
ρ	Resistivity	ohm·meter	Ω·m
α	Angle	degree	°
ψ	Slot spacing	electrical degree	°
m	Number of phases		
n	Slots/phase/pole		
p	Pole span	electrical degree	°
X	Reactance	ohm	Ω
L	Inductance	henry	H
Z	Impedance	ohm	Ω
δ	Torque or power angle	electrical degree	°
a	Turns ratio		
pf	Power factor		
s	Slip		
α	Transformation ratio		
C	Capacitance	farad	F

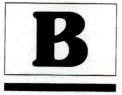

Transformer Efficiency from Tests

When a transformer delivers power to a load, the ac source connected to the primary will supply this power in addition to the transformer's own power losses. Obviously, the power output of the transformer is less than its power input, which implies the efficiency of the transformer is less than 100%. As discussed in Chapter 7, transformer efficiency is determined by the equation,

$$\eta = \frac{\text{power out}}{\text{power out} + \text{copper loss} + \text{core loss}} \times 100\%$$

The core loss is practically constant for all loading conditions, including no load, and is obtained from a no-load test. The copper loss is proportional to the square of the load current. Therefore, knowing the full load copper loss from a short-circuit test, it is a simple matter to calculate this loss at any fraction of the full-load current. Thus the efficiency versus output kVA (or kW) can be plotted for a transformer (knowing the results of the open-circuit and short-circuit tests) over its entire operating range for a particular power factor.

EXAMPLE B.1

Using the data of Example 7.6, namely: A 10-kVA 2400/240-V 60-Hz transformer was tested with the following results:

$$\text{Input during short-circuit test} = 340 \text{ W}$$

$$\text{Input during open-circuit test} = 168 \text{ W}$$

Determine the efficiency of this transformer versus load power over the entire load range up to 130% overload. Use a load PF = 0.8.

Solution   Because of the repeated calculations required, use will be made of a computer-aided solution. The program given below is written in FORTRAN. Using this program, the printout is shown in Table B.1 while Figure B.1 shows these results graphically.

As can be seen, specific transformer data must be typed in. This makes the program generally applicable, and it can readily be used for transformers of various sizes. The results of Figure B.1 show that the transformer efficiency is at its maximum when the fixed losses (the core loss) are equal to the variable losses (the copper loss). Where this maximum occurs is usually determined by user specifications, which must be known prior to implementing the specific transformer design. It should be realized that like other electrical machines, transformers seldom run at full load continually. Therefore, the point of maximum efficiency never occurs at full load but generally in the range 60 to 80% of rated value.

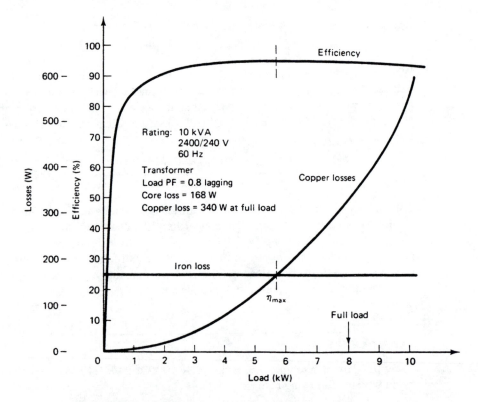

FIGURE B.1.   Graphical representation of results from computer program.

```
      REAL KVA,LOAD
C
C     Transformer Efficiency Performance
C
      KVA=10.
      CL=168.
      CUL=340.
      PF=0.8
C     CL=core loss; CUL=copper loss at full load
C     KVA=rated kVA; PF=power factor; PL=% of F.L.
      WRITE(2,3)
   3  FORMAT(7X,29H Transformer Efficiency Curve /)
      WRITE(2,4)
   4  FORMAT(8X,3H kW,4X,8H cu loss,3x,11H efficiency )
      DO 6 I=1,13
      AI=I
      PL=AI/10.
      PCUL=PL*PL*CUL
      LOAD=KVA*PL*PF
      EFF=LOAD/(LOAD+(CL+PCUL)/1.E3)*100.
      WRITE(2,8)  LOAD,PCUL,EFF
   6  CONTINUE
   8  FORMAT(5X,F6.1,5X,F6.1,7X,F6.2)
      STOP
      END
```

TABLE B.1.   **Transformer efficiency curve**

kW	cu loss	efficiency
0.8	3.4	82.36
1.6	13.6	89.81
2.4	30.6	92.36
3.2	54.4	93.50
4.0	85.0	94.05
4.8	122.4	94.30
5.6	166.6	94.36
6.4	217.6	94.32
7.2	275.4	94.20
8.0	340.0	94.03
8.8	411.4	93.82
9.6	489.6	93.59
10.4	574.6	93.34

The Rotating Magnetic Field

When a three-phase winding is connected to a three-phase power supply, currents will flow in each of the phase windings. They will be displaced from each other by 120 electrical degrees, as shown in Figure C.1, assuming an *abc* phase sequence. Next, the following convention will be adopted. With reference to Figure C.2, phase winding *a* when carrying positive current will produce a resulting MMF in the positive direction of its magnetic axis, as determined by the right-hand rule. Furthermore, the positive current is assumed to enter the winding at *a* and to be directed into the page, as indicated by $\oplus$. When the current i_a becomes negative, the current direction reverses and the MMF direction will also be opposite. Similar reasoning is applied to phase windings *b* and *c*. Referring to Figure C.1, four instants in time are selected, and it will be shown that the resultant MMF, $\mathcal{F}_t$, is constant in magnitude and revolves the distance covered by one pole pitch or 180 electrical degrees in one-half cycle.

At time t = t₁. In Figure C.1, i_a is positive maximum, and i_b and i_c are negative and equal to one-half their respective maxima. Therefore, the MMFs due to these currents will have corresponding magnitudes and directions, as determined by the right-hand rule and our adopted convention, as indicated in Figure C.3a. Combining the three component MMFs yields a resultant MMF $\mathcal{F}_t$, which is 1.5 times the maximum MMF $\mathcal{F}_m$ per phase and directed, according to our assumptions made, horizontally from right to left.

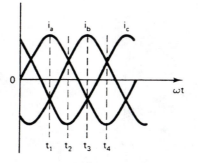

FIGURE C.1. **Three-phase currents in stator winding.**

At time t = t₂. At this instant in time, i_a and i_b are positive and equal to one-half their respective maximum values, and i_c is maximum but negative. The MMFs will have corresponding directions and magnitudes and when combined, the resultant MMF is again 1.5 times the maximum MMF per phase but is rotated 60° in a clockwise direction compared to that of $\mathcal{F}_t$ at $t = t_1$ (see Figure C.3b). Note the MMF phasor $\mathcal{F}_t$ rotated 60 mechanical degrees in space which follows the time increment from t_1 to t_2, which equals 60 electrical degrees.

At time t = t₃. i_b is positive maximum and i_a and i_c are negative with a magnitude equal to half their maximum values. Proceeding as before, the individual MMFs proportional to the current magnitudes and directed as dictated by the current directions are combined. As before, the resultant MMF is $\mathcal{F}_t$, but it has rotated an additional 60° in the clockwise direction, as illustrated in Figure C.3c. This corresponds to the time increment from t_2 to t_3, which is also 60°.

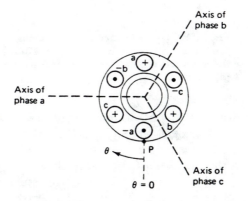

FIGURE C.2. **Simplified two-pole three-phase stator winding showing polarities for positive phase currents.**

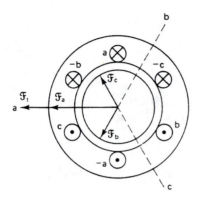

(a)

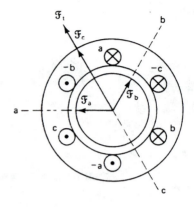

(b)

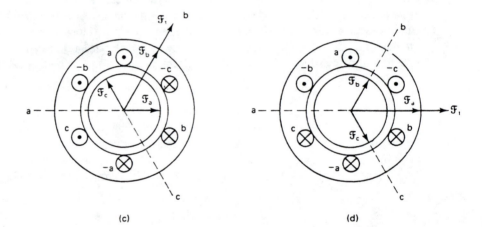

(c)

(d)

FIGURE C.3.   Illustration of rotating magnetic field for the four time instants indicated in Figure C.1, for a two-pole three-phase stator winding: (a) $t = t_1$; (b) $t = t_2$; (c) $t = t_3$; (d) $t = t_4$.

At time $t = t_4$. At this instant in time, 60 electrical degrees later than t_3, current i_a is negative maximum, while i_b and i_c are one-half their respective maximum value and positive. Determining the corresponding MMF directions and magnitudes, they are combined as shown in Figure C.3d. The resulting MMF is again $\mathcal{F}_t$ and rotated an additional 60° as compared to its direction at $t = t_3$. The resulting MMF is thus seen to have rotated 180° in space, or one pole pitch. This has happened in the time that the current i_a changed from positive maximum to negative maximum, representing one-half cycle of the supply voltage. Therefore, the field will revolve a distance covered by two poles for each cycle of the supply frequency. Thus, continuing for a complete cycle of the supply frequency it will be seen that the resultant MMF rotates a full cycle, for a two-pole machine. For a four-pole machine, the revolving field would have rotated only one-half a revolution for each full cycle of the supply frequency, and so on. This implies that the speed of the rotating field is inversely proportional to the number of pole pairs. Also, as discussed, for a two-pole stator the speed of rotation is 1 r/s or 60 r/min, for a frequency of 1 Hz, two rotations for a frequency of 2 Hz, or 60 rotations for 60 Hz. The speed is thus proportional to the frequency of the supply, f. Expressing this by a formula gives what is known as the synchronous speed n_s, of the rotating magnetic field, namely,

$$n_s = \frac{f}{P/2} \text{ r/s} = \frac{120 f}{P} \text{ r/min} \tag{C.1}$$

C.1   Analytical Analysis

In the graphical analysis each phase was excited by an alternating current which varies sinusoidally with time. For a balanced supply with abc sequence and taking i_a as a reference, the instantaneous currents are

$$i_a = I_m \sin \omega t \tag{C.2}$$

$$i_b = I_m \sin(\omega t - 120°) \tag{C.3}$$

$$i_c = I_m \sin(\omega t - 240°) \tag{C.4}$$

The corresponding components of the MMFs therefore vary sinusoidally with time, since they are proportional to the currents. Each component by itself is stationary, although pulsating sinusoidally with its maximum located on the magnetic axis of each phase. In other words, each component represents an oscillating vector along the magnetic axis of its phase. Its length is proportional to the instantaneous current value and its direction determined by the respective phase current polarity. This is clearly evident from Figure C.3. The resultant MMF is simply the sum of the three MMF components. Therefore, the MMF contribution of each phase at any point along the air gap periphery is

$$\mathcal{F}_a = \mathcal{F}_{max} \sin \theta \tag{C.5}$$

$$\mathcal{F}_b = \mathcal{F}_{max} \sin(\theta - 120°) \tag{C.6}$$

and

$$\mathcal{F}_c = \mathcal{F}_{max} \sin(\theta - 240°) \tag{C.7}$$

If point P in Figure C.2 represents a point of zero MMF for phase a, then for a point θ electrical degrees from P the MMF due to phase a equals

$$\mathcal{F}_a = \mathcal{F}_{max} \sin \theta = \mathcal{F}_m \sin \omega t \sin \theta \tag{C.8}$$

Similarly, the MMF due to phase b is

$$\mathcal{F}_b = \mathcal{F}_m \sin(\omega t - 120°) \sin(\theta - 120°) \tag{C.9}$$

and the MMF due to phase c is

$$\mathcal{F}_c = \mathcal{F}_m \sin(\omega t - 240°) \sin(\theta - 240°) \tag{C.10}$$

where $\mathcal{F}_m$ is the maximum value at a particular instant of time of the amplitude of $\mathcal{F}_{max}$, since the MMF $\mathcal{F}_{max}$ in itself varies with time in accordance with the current variation. Hence the resultant MMF at a point θ degrees from P becomes

$$\mathcal{F}_t = \mathcal{F}_a + \mathcal{F}_b + \mathcal{F}_c$$

$$= \mathcal{F}_m [\sin \omega t \sin \theta + \sin(\omega t - 120°) \sin(\theta - 120°)$$

$$+ \sin(\omega t - 240°) \sin(\theta - 240°)]$$

$$= \tfrac{3}{2} \mathcal{F}_m (\cos \omega t \cos \theta + \sin \omega t \sin \theta)$$

or

$$\mathcal{F}_t = \frac{3}{2} \mathcal{F}_m \cos(\omega t - \theta) \tag{C.11}$$

Equation C.11 shows that the magnitude of $\mathcal{F}_t$ is a constant and equal to $1.5 \mathcal{F}_m$. Furthermore, the total air gap MMF $\mathcal{F}_t$ rotates at the angular velocity of ω radians per second and θ is the mechanical angle along the periphery, as measured from point P.

For $t = 1/f$, that is, the period of the supply voltage, $\theta = 2\pi$ rad or 360°, which indicates the maximum of the resultant MMF has rotated through two pole pitches in one cycle of the supply voltage. If the machine has P poles, the resultant magnetic MMF rotates through 2/P revolutions in one cycle of the supply. Therefore, the speed of the rotating magnetic flux, n_s, equals

$$n_s = \frac{2f}{P} \text{ r/s} \quad \text{or} \quad n_s = \frac{120f}{P} \text{ r/min}$$

consistent with Eq. C.1. It is worth noting that by reversing the phase sequence of any of the two currents in Figure C.1 results in an MMF rotating opposite to the one discussed. This is readily verified. In practice, this is the procedure to reverse the direction of rotation of a three-phase motor, whereby any two of the motor supply leads are interchanged at the motor terminal box. The ease by which it is possible to reverse the direction of rotation constitutes one of the advantages of three-phase motors.

The following example illustrates the rotating magnetic field.

EXAMPLE C.1

a. Write a program to calculate the resultant MMF from Eqs. C.8 to C.10 and compare this result with that obtained by using Eq. C.11. For this take an arbitrary value of $\mathcal{F}_m = 10$, and vary ωt from 0° to 360° in steps of 10°.

b. Repeat the calculations for various space angles θ, say 30° intervals.

c. Check the rotation of the field by monitoring the maximum values at different $\theta = \omega t$.

Solution **a., b.** The program to calculate the individual MMF contributions due to the three phases $\mathcal{F}_1$, $\mathcal{F}_2$, and $\mathcal{F}_3$, their sum $\mathcal{F}_t$ representing the total MMF, and the total MMF as calculated by the use of Eq. C.11 and designated as $\mathcal{F}_4$ is given below.

```
C   Program calculates revolving air-gap mmf, Ft
C
    PI=3.14159
    THETA=PI/2.
    FM=10.
  6 WRITE(2,4)
  4 FORMAT(/12X,3H wt,6X,11H Sum of F's,6X,3H Ft,6X,12H Space angle)
    DO 3 I=1,37
    AI=I-1
    PHI=AI*10.
    WT=PHI*PI/180.
    C1=2.*PI/3.
    C2=2.*C1
    F1=FM*SIN(THETA)*SIN(WT)
    F2=FM*SIN(THETA-C1)*SIN(WT-C1)
    F3=FM*SIN(THETA-C2)*SIN(WT-C2)
    FT=F1+F2+F3
    F4=(3.*FM/2.)*COS(THETA-WT)
C   FT is the sum of the component mmf's = F4
    WRITE(2,5) PHI,FT,F4,THETA
  3 CONTINUE
  5 FORMAT(5X,4E13.3)
    THETA=THETA+PI/6.
    IF(THETA.LE.PI) GO TO 6
    STOP
    END
```

Because of the lengthy printout the results are represented graphically in Figure C.4 for three different time intervals.

c. As can be seen, the MMF distribution is sinusoidal and its peak value moves linearly from $\theta = \omega t = 90°$, through 120° to 150° in a time interval corresponding to one-sixth of a period of the supply frequency, for the two-pole machine.

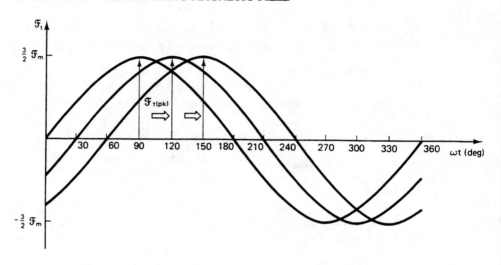

FIGURE C.4. Sinusoidal distribution of the resultant air gap MMF due to three-phase stator currents at three different time intervals (see Example C.1).

Induction Motor Characteristics

Example 8.3 showed us how to calculate the induction motor performance at a specific slip value given the equivalent circuit diagram for the machine. In Section 8.10 we discuss the necessary tests to be performed on the motor in order to determine the parameter values for the equivalent circuit. We may recall that it is necessary to carry out a blocked-rotor test, a no-load test, and a resistance measurement on the stator winding.

As can be appreciated, when the complete performance curves for the motor are required over an extended range of slip values, it becomes desirable to resort to computer computations. This has been done here. The computer program given below calculates the performance data for the induction motor in Example 8.3, for a slip range that can be considered its normal operating range.

The obtained data are also included, it shows for example the calculations at a slip $s = 0.028$, the value used in Example 8.3, which represents full load for this particular motor.

Some interesting observations can be made from the results, verifying the theory presented in Chapter 8. For small slip values (i.e., a lightly loaded machine), the line current is fairly large, due to the magnetizing current. As we can see, this current is an appreciable large component when compared with the full-load current. With increased loading (larger slip values), the power increases rapidly and the efficiency goes up, reaching a maximum, and then starts to decrease again. Note that the efficiency is not maximum at full load ($s = 0.28$) but somewhat below rated value. As a matter of fact, it peaks at about 75% of full load. The reason for this is similar to that for the transformer, as discussed in Appendix B.

The program can readily be extended to calculate the motor's performance over the entire slip range or modified to suit other specific motor data. From the results the various performance curves can be plotted for graphical representation. However, this will be left for the reader to pursue.

```
C       Calculate Induction Motor Characteristics
C       Simplified circuit model is used
C       Calculations are restricted to normal operating range
C
C       Input motor data on a per phase basis
C       Vin=input voltage, Rs=stator R, Rp = R', Xs=stator X
C       Xp=X', Xm=magnetizing X, Prot=rotational loss
C
        COMPLEX Zp,Ip,Im,Il
        REAL ns,nr
        PI=3.141593
        Vin=127.
        Rs=0.344
        Rp=0.147
        Xs=0.498
        Xp=0.224
        Xm=12.6
        Prot=262.
        ns=1200.
        WRITE(2,4)
    4   FORMAT(/8x,2H s,4x, 7H I-line,7x,3H PF,5x,3H nr, 5x,
    1   7H Torque,6x,3H hp, 6x,4H Eff /)
C
C       s=slip, IM-I magnetizing, Il=line current, nr=rotor speed
C       CIl=magnitude of line current, ns=synchronous speed
C       CIp=magnitude of I'
C
        DO 6 I=2,32,2
          A=I
          s=A/1000.
          Zp=CMPLX((Rs+Rp/s),(Xs+Xp))
          Ip=Vin/Zp
          CIp=CABS(Ip)
          Im=Vin/CMPLX(0.0,Xm)
          Il=Ip+Im
          CIl=CABS(Il)
          AI=AIMAG(Il)
          RI=REAL(Il)
          THETA=ATAN2(AI,RI)
          PF=COS(THETA)
```

```
      nr=(1.-s)*ns
      wr=2.*PI*nr/60.
      RPI=3.*CIp*CIp*Rp/s
      RPD=(1.-s)*RPI
      Pout=RPD-Prot
      Tout=Pout/wr
      hp=Pout/746.
      Pin=3.*Vin*CIl*PF
      EFF=(Pout/Pin)*100.
      WRITE(2,8)  s,CIl,PF,nr,Tout,hp,EFF
6     CONTINUE
8     FORMAT(1X,F10.3,2F10.2,4F10.1)
      STOP
      END
```

s	I-line	PF	nr	Torque	hp	Eff
.002	10.24	.17	1197.6	3.1	0.5	59.3
.004	10.71	.32	1195.2	8.2	1.4	78.6
.006	11.43	.45	1192.8	13.2	2.2	84.6
.008	12.36	.55	1190.4	18.1	3.0	87.2
.010	13.45	.63	1188.0	22.8	3.8	88.6
.012	14.65	.69	1185.6	27.5	4.6	89.3
.014	15.93	.73	1183.2	32.1	5.3	89.6
.016	17.27	.77	1180.8	36.6	6.1	89.7
.018	18.66	.79	1178.4	41.0	6.8	89.6
.020	20.07	.82	1176.0	45.3	7.5	89.4
.022	21.49	.83	1173.6	49.4	8.1	89.2
.024	22.93	.85	1171.2	53.5	8.8	88.9
.026	24.38	.86	1168.8	57.5	9.4	88.5
.028	25.82	.86	1166.4	61.4	10.1	88.1
.030	27.26	.87	1164.0	65.2	10.7	87.7
.032	28.70	.88	1161.6	68.9	11.2	87.3

References

1. O. I. Elgard, *Basic Electric Power Engineering*, Addison-Wesley, Reading, MA, 1977.

2. I. L. Kosow, *Electric Machinery and Transformers*, 2nd Ed., Prentice-Hall, Englewood Cliffs, NJ, 1991.

3. V. Del Toro, *Electromechanical Devices for Energy Conversion and Control Systems*, Prentice-Hall, Englewood Cliffs, NJ, 1968.

4. P. F. Ryff, D. Platnick, and J. A. Karnas, *Electrical Machines and Transformers— Principles and Applications*, Prentice-Hall, Englewood Cliffs, NJ, 1987.

5. S. J. Chapman, *Electric Machinery Fundamentals*, 2nd Ed., McGraw-Hill, 1991.

6. J. F. Lindsay and M. H. Rashid, *Electromechanics and Electrical Machinery*, Prentice-Hall, 1986.

7. R. Ramshaw and R. G. van Heeswÿk, *Energy Conversion, Electric Motors and Generators*, Saunders College Publishing, 1989.

8. P. C. Sen, *Principles of Electric Machines and Power Electronics*, John Wiley and Sons, 1989.

9. J. D. Edwards, *Electrical Machines, An introduction to principles and characteristics*, 2nd Ed., MacMillan Publishing Company, 1986.

10. Dino Zorbas, *Principles, Applications and Control Schematics*, West Publishing Company, 1989.

Answers To Problems

CHAPTER 1

1.1 2387   **1.2** 1084 At   **1.3** 3.08×10^{-3} H/m; 2.16×10^{-3} H/m; 1.67×10^{-3} H/m; 0.604×10^{-3} H/m   **1.4** 2450; 1720; 1330; 480   **1.5** 1.73A   **1.6** 84A; 386%   **1.7** (a) 432 μWb; (b) 1.10A   **1.8** (a) 4mWb, 0.65A; 6mWb, 1.12A; 8mWb, 1.26A; 10mWb, 2.7A; 12mWb, 6.0A; (b) 4mWb, 2.64A; 6mWb, 4.10A; 8mWb, 5.73A; 10mWb, 7.68A; 12mWb, 11.95A   **1.9** 2.29×10^3N; 2870At.   **1.10** (a) 1.04A; (b) 51N; (c) 152mA   **1.11** 0.44mWb   **1.12** 0.62 mWb   **1.13** 2285At   **1.14** 536W; 324W at 60Hz; 224W; 56W at 25Hz   **1.15** 2.74A   **1.16** 6.5A   **1.17** 1.69A   **1.18** 0.42T

CHAPTER 2

2.1 2.4mA from a to b **2.2** 9.6×10^{-5}N to the left   **2.3** (a) 76V; (b) 228V; 80A; (c) 18.24 kW Lap and wave wound   **2.4** 4.5N   **2.5** (a) 698.9N; (b) 83.9N·m; (c) 14050W; 234.2V   **2.6** 83 Conductors   **2.7** (a) 63.06; 6.6; (b) 173.9V; (c) 463.1N·m; (d) 41.74 kW   **2.8** (a) 13.2; 126.1; (b) 347.4V; (c) 463.1N·m; (d) 41.74kW   **2.9** (a) 45°; (b) 0.432×10^{-5} N·m

CHAPTER 3

3.1 166.7V   **3.2** 109.4V   **3.3** 46.2A   **3.4** Reduced from 160 to 95Ω   **3.5** 91.1%   **3.6** 1300At/coil   **3.7** 8.5kW   **3.8** 3.2 turns   **3.9** 2.4%   **3.10** (a) 197.3V; (b) 202.2V

CHAPTER 4

4.1 976 r/min   **4.2** 975 r/min   **4.3** (a) 2083W; (b) 1843W; (c) 14.7 N·m; (d) 78.8%   **4.4** R = 0.628Ω   **4.5** (a) 2153W; (b) 4.89%   **4.6** 82.8%   **4.7** 3.67Ω   **4.8** 3.21Ω   **4.9** (a) 2.45Ω; (b) 137.5V   **4.10** (a) 89.1 N·m; (b) 237 N·m; (c) 1326 r/min   **4.11** 2.8Ω, 66.7V **4.12** (a) 55866kW; (b) 87.2%; (c) 8201W   **4.13** 580 r/min   **4.14** (a) 87.1A, 198.2V; (b) 105.5 r/min or 1.98 m/s; (c) 117 r/min, 2.2 m/s   **4.15** 84.4%   **4.16** 2.78Ω, 50.6%   **4.17** (a) 382 r/min; (b) 42 N·m; (c) 1848W or 2.5 hp

CHAPTER 5

5.1 900 r/min   **5.2** 27 1/3 to 65 1/3 Hz   **5.3** 262.4A   **5.4** 236.2A   **5.5** (a) 4839W; (b) 14.52kW   **5.6** 250kVA   **5.7** 2875V   **5.8** 4980V in Y; 2875V when Δ-connected. **5.9** (a) 60 Hz; (b) $e_A = 45.2 \sin 377t$ V, $e_C = 45.2 \sin(377t - 120°)$V, $e_B = 45.2 \sin(377t + 120°)$V; (c) 32∠0V, 32∠−120°V, 32∠+120°V   **5.10** 55.4∠30°V, 55.4∠−90°V, 55.4∠150°V **5.12** 12 el., degrees, 0.957   **5.13** (a) 0.951; (b) 0.910   **5.14** 13.8 kV   **5.15** (a) 6; (b) 3 1/3 mech. degrees; (c) 0.956

CHAPTER 6

6.1 2387 V/ϕ **6.2** 112A   **6.3** 169A   **6.4** 286V   **6.5** 614V   **6.6** 37.3%   **6.7** (a) 63.0%; (b) −18.3%   **6.8** (a) 21.0Ω; (b) 81.5% (c) 24.2°; (d) 749.83 kW; (e) 1000 kVA **6.9** 93.6%   **6.10** 24.5%   **6.11** (a) 60 poles; (b) 108 MW; (c) 4518A; (d) 111,340 kW; (e) 8860 kN·m   **6.12** 2.77Ω/phase, 166.2V/phase   **6.13** (a) 15.2°; (b) 26.6°; (c) 16.56kV L-L   **6.14** −20.1%   **6.15** 42.2%   **6.16** 2126V

CHAPTER 7

7.1 (a) 5.63 mWb; (b) 1200V   **7.2** 12.5A, 125A   **7.3** (a) 1.458Ω, 0.091Ω   **7.4** (a) 327V; (b) 0.28   **7.5** $Z_{EH} = 11.1$Ω, 4512V   **7.6** (a) 97.2%; (b) 96.6%; (c) 97.21%   **7.7** 93.87%   **7.8** (a) $Z_{EP} = 33.1$Ω; (b) 4.5%   **7.9** $Z_{EH} = 0.437$Ω; (a) 97.74% at rated, 97.59% 1/2-rated; (b) 3.06%; (c) 97.81%   **7.10** 93.4%   **7.11** 97.53%   **7.12** (a) 550V, $I_S = 18.18$A; (b) 4.84Ω   **7.13** 17.8A   **7.14** 89.94%   **7.15** 2.22%   **7.16** 93.7%   **7.17** 57.83 kVA   **7.18** 1375 kVA   **7.19** $I_{LP} = 1.88$A, $V_{LP} = 6894$V   **7.20** (a) 47.4; (b) 15.58kW/transformer; (c) 69.0A; (d) $I_{LP} = 2.52$A

CHAPTER 8

8.1 564 r/min   **8.2** (a) 0.044; (b) 2.64Hz; (c) 40 r/min   **8.3** (a) 6 poles; (b) 5%; (c) 3 Hz; (d) i) 60 r/min, ii) 1140 r/min; iii) ϕ r/min; (e) 60 Hz   **8.4** (a) 12 poles; (b) 7%   **8.5** 0.033 $n_r = 1160$ r/min   **8.7** V/phase (rotor) = 7.04V   **8.8** (a) V/phase (rotor) = 110V, $f_R = 60$ Hz; (b) 4.4V, 2.4 Hz; (c) 6 poles, $s = 1.67$, V phase (rotor) = 183.7, $f_R = 100$ Hz   **8.10** (a) $I_L = 42.4$A; (b) 106.8 N·m, 17.45 hp; (c) 89.5%   **8.11** $s \approx 3.68\%$ assuming $R_{st} \approx 0$ **8.12** $R_{DC} = 1.042\Omega$/phase, R_{eff}/phase = 1.250Ω **8.13** (a) $148.4\angle - 26.4$A, P.F. = 0.896; (b) 1007 N·m, 123.4 hp; (c) 90.9%   **8.14** (a) 0.116; (b) 796 r/min; (c) $I_L \approx 401$A since $I_m << I'_{R_mT}$, $T_{mT} = 2026$ N·m   **8.15** $I_L \approx I'_{R_{st}} + I_m = 581$A, $T_{st} = 497$ N·m   **8.16** (a) 0.078; (b) 0.853; (c) 214.6 N·m; (d) 89.7%   **8.17** (a) $Z_{BR} = 1.48\Omega$, $R'_r = 0.30\Omega$, $x'_e = 1.34\Omega$; (b) $I_L = 18.2\angle - 30°$; (c) 92.7%   **8.18** (a) $R'_r = 0.142\Omega$ selected; (b) $S_{mT} = 0.145$, $n_r = 1539$ r/min, 86.8 N·m; (c) 33.82 N·m   **8.19** 11.8 N·m   **8.20** (a) 52 N·m; (b) 83 N·m, $V_{st} = 379$V; (c) 87.3A; (d) $I_{st} = 32$A, 110V   **8.21** (a) 1 p.u.; (b) $T_{st} = 0.33$, $I_{st} = 0.58$ p.u.; (c) $T_{st} = 0.36$ p.u., $I_{st} = 0.6$ p.u.

CHAPTER 9

9.1 0.0313 N·m, 0.023 lb.ft, 4.43 oz.in   **9.2** 57.8%   **9.3** $C = 216.5\mu$F, $I_L = 21.8$A **9.4** V/turn = 0.697   **9.5** 144μF start; 9μF run; 720V   **9.6** 1.75Ω **9.7** (a) 34V; (b) 0.35 N·m; (c) $n \approx 250$ rad/s; (d) 27.2W; (e) 0.26 N·m   **9.8** (a) 39V; (b) 160 rad/s; (c) 250 rad/s, 0.23 N·m, 57.5W; (d) 39V to 61V   **9.9** (a) 2969 r/min; (b) 30.9% without R, 20% with R   **9.10** (a) 925 r/min; (b) 0.355 N·m   **9.11** 1.53Ω **9.12** 35.5 rad/s full load, 43.7 rad/s on noload   **9.13** 142W   **9.14** 50 in.lb, 20A   **9.15** 4 phases on stator, 5 poles on rotor; clockwise $ABCDA$

CHAPTER 10

10.1 (a) Generator 10 poles; (b) Motor 24 poles; (c) 300 r/min   **10.2** Generator 10 poles, Motor 12 poles   **10.3** (a) 600 r/min; (b) speed remains constant   **10.4** $X_s = 10.2\Omega$, $\delta = -6.3°$, $I_f = 155$A   **10.5** (a) 97.4%; (b) 716 hp   **10.6** $\delta = 20.6°$, $I_A = 247.5\angle 21°$ **10.7** 2356 kVA total, $I_S = 2957$, overloaded by about 18%   **10.8** 9.76%   **10.9** 0.755Ω **10.10** (a) 1.434 MW; (b) 1.122 MW   **10.11** Operates at lagging P.F.   **10.12** P.F. becomes leading   **10.13** 16,687W, P.F. = 0.644   **10.14** (a) 44.4°; (b) $534\angle - 11°$A   **10.15** (a) 44.3V; (b) 84.9° leading   **10.16** P.F. = 0.965 leading   **10.17** (a) S.C. kVAR = 761, 960kW total   **10.18** (b) kVA S.M. = 1189; P.F. S.M = 0.348

Index